The Palgrave Macmillan Animal Ethics Series

Series Editors
Andrew Linzey
Oxford Centre for Animal Ethics
Oxford, UK

Clair Linzey
Oxford Centre for Animal Ethics
Oxford, UK

In recent years, there has been a growing interest in the ethics of our treatment of animals. Philosophers have led the way, and now a range of other scholars have followed from historians to social scientists. From being a marginal issue, animals have become an emerging issue in ethics and in multidisciplinary inquiry. This series will explore the challenges that Animal Ethics poses, both conceptually and practically, to traditional understandings of human-animal relations. Specifically, the Series will:

- provide a range of key introductory and advanced texts that map out ethical positions on animals;
- publish pioneering work written by new, as well as accomplished, scholars;
- produce texts from a variety of disciplines that are multidisciplinary in character or have multidisciplinary relevance.

For further information or to submit a proposal for consideration, please contact Amy Invernizzi, amy.invernizzi@palgrave-usa.com.

Doris Schneeberger

Envisioning a Better Future for Nonhuman Animals

Towards Future Animal Rights Declarations

palgrave
macmillan

Doris Schneeberger
Vienna University of Economics and Business
Vienna, Austria

ISSN 2634-6672 ISSN 2634-6680 (electronic)
The Palgrave Macmillan Animal Ethics Series
ISBN 978-3-031-75331-2 ISBN 978-3-031-75332-9 (eBook)
https://doi.org/10.1007/978-3-031-75332-9

This Palgrave Macmillan imprint is published by the registered company Springer Nature Switzerland AG.
The registered company address is: Gewerbestrasse 11, 6330 Cham, Switzerland

Because "being a bystander to suffering is not an option." (Singer 2016, 40)

Acknowledgements and Foreword

I would like to thank all those who have made it possible for me to write this book. My thanks in the first instance to Robert Kogler, for coming up with the idea of my writing my philosophical PhD thesis on a possible future universal declaration of animal rights. I would also like to thank my parents for their support (financial and otherwise), which has enabled me to take the time necessary to devote myself fully to this project. I would like to thank Julian Abernethy for cheering me on and for his boundless support. My sincere thanks are due too to the scholars who have accompanied and supported me over the course of years, particularly Heinrich Ganthaler and Otto Neumaier, who supervised the writing of the philosophical PhD thesis on which this book is based, as well as Anne Siegetsleiter and Markus Hametner, who provided helpful feedback on earlier versions of this book. I would moreover like to thank Tom Beauchamp, Peter Singer, Nico Müller, Markus Wild, Will Kymlicka, and Sue Donaldson for welcoming me as a guest researcher and for supporting the development of this book. I deeply thank my proofreader, Alex Schott Fairley, for his outstanding work and kindness. I extend my heartfelt thanks to Amy Invernizzi for her thoughtful guidance during the publication process and to the two anonymous reviewers for their insightful and nuanced comments on an earlier version of this book. My thanks goes to all the people who have reflected on animal ethical questions in discussions and conversations, and particularly to those who have devoted their intellectual lives to animal ethical questions. At this point in time, we might yet be unable to comprehend the importance of this intellectual work in fostering the future moral growth of society. My thanks are also due to all nonhuman and human animal beings who are and have been my friends, as well as to all the people who have served as role models for me, who have believed in me and continue to believe in me, and the nonhuman animals in my life who have brought with them so much joy and love. Thank you to those who work to make this world a better place. Thank you to those who stand up for and stand in solidarity with the powerless and defenceless. Thank you to you, for reading these words: it is up to us to shape the future, it is up to us to commit to what really matters. It is up to us to break new ground, and to never stop working on ourselves. It is up to us to use our lives to make this world a better place where there is less suffering and injustice.[1]

[1] There are various definitions and theories of justice. I use the term "justice" here in the common sense of referring to "the quality of being just, impartial, or fair" (*Merriam-Webster* online dictionary n.d.d.).

Our Privilege Is Our Responsibility

Most likely, you—like me—are a resident of one of the richest countries in the world. Humans living in affluent countries have won a precious ticket in the lottery of life. On the one hand, this lottery ticket is a privilege that allows us to enjoy an exceptional quality of life; on the other, this lottery ticket brings with it a responsibility to use this privilege to advocate for those who have drawn a less advantageous lottery ticket; to advocate for their most important goods to be protected and their interests to be taken into account fairly. Thank you to all those who, firstly, perceive the imperative that comes with this privilege, and secondly, have been trying to make a difference out of this awareness. I hope that humanity will continue its moral growth in the anti-speciesist dimension, so that we can reach new heights of morality beyond even the imagination of the animal ethicists of my lifetime.

Series Editors' Preface

This is a new book series for a new field of inquiry: Animal Ethics.

In recent years, there has been a growing interest in the ethics of our treatment of animals. Philosophers have led the way, and now a range of other scholars have followed from historians to social scientists. From being a marginal issue, animals have become an emerging issue in ethics and in multidisciplinary inquiry.

In addition, a rethink of the status of animals has been fuelled by a range of scientific investigations which have revealed the complexity of animal sentiency, cognition, and awareness. The ethical implications of this new knowledge have yet to be properly evaluated, but it is becoming clear that the old view that animals are mere things, tools, machines, or commodities cannot be sustained ethically.

But it is not only philosophy and science that are putting animals on the agenda. Increasingly, in Europe and the United States, animals are becoming a political issue as political parties vie for the "green" and "animal" vote. In turn, political scientists are beginning to look again at the history of political thought in relation to animals, and historians are beginning to revisit the political history of animal protection.

As animals grow as an issue of importance, so there have been more collaborative academic ventures leading to conference volumes, special journal issues, and indeed new academic animal journals as well. Moreover, we have witnessed the growth of academic courses, as well as university posts, in Animal Ethics, Animal Welfare, Animal Rights, Animal Law, Animals and Philosophy, Human-Animal Studies, Critical Animal Studies,

Animals and Society, Animals in Literature, Animals and Religion—tangible signs that a new academic discipline is emerging.

"Animal Ethics" is the new term for the academic exploration of the moral status of the non-human—an exploration that explicitly involves a focus on what we owe animals morally, and which also helps us to understand the influences—social, legal, cultural, religious, and political—that legitimate animal abuse. This series explores the challenges that Animal Ethics poses, both conceptually and practically, to traditional understandings of human-animal relations.

The series is needed for three reasons: (i) to provide the texts that will service the new university courses on animals; (ii) to support the increasing number of students studying and academics researching in animal-related fields; and (iii) because there is currently no book series that is a focus for multidisciplinary research in the field.

Specifically, the series will

- provide a range of key introductory and advanced texts that map out ethical positions on animals;
- publish pioneering work written by new, as well as accomplished, scholars; and
- produce texts from a variety of disciplines that are multidisciplinary in character or have multidisciplinary relevance.

The new Palgrave Macmillan Series on Animal Ethics is the result of a unique partnership between Palgrave Macmillan and the Ferrater Mora Oxford Centre for Animal Ethics. The series is an integral part of the mission of the Centre to put animals on the intellectual agenda by facilitating academic research and publication. The series is also a natural complement to one of the Centre's other major projects, the *Journal of Animal Ethics.* The Centre is an independent "think tank" for the advancement of progressive thought about animals and is the first Centre of its kind in the world. It aims to demonstrate rigorous intellectual enquiry and the highest standards of scholarship. It strives to be a world-class centre of academic excellence in its field.

We invite academics to visit the Centre's website (www.oxfordanimalethics.com) and to contact us with new book proposals for the series.

Oxford Centre for Animal Ethics
Oxford, UK

Andrew Linzey
Clair Linzey

About the Book

Humans are granted animal rights which are enshrined in the Universal Declaration of Human Rights. Should nonhuman animals have rights, too? And if so, which animals should have which rights? And why? This book contributes to the interdisciplinary and philosophical discourse around these questions. It provides an overview of central debates in the field of animal ethics, discusses the status quo with regard to how humans currently treat nonhuman animals, and integrates biological and ethological knowledge into ethical discussions. The main part of the book consists of normative reflections on and descriptive accounts of 12 potential universal nonhuman animal rights. This book provides argumentation as to why certain rights should be granted to certain nonhuman animals as well. It is hoped that it will prove invaluable to students, scholars, and practitioners interested in nonhuman animal rights, inviting readers to reflect on interdisciplinary aspects of nonhuman animal rights, whilst furthering scholarly inquiry and advocacy in the social movement working for nonhuman animal rights.

Contents

CHAPTER 1

About this Book

1.1 How this Book Came into Existence

This book is a revised version of my philosophical PhD thesis, which I submitted in 2016 at the University of Salzburg. I wrote my thesis in German between 2013 and 2016, and quoted numerous German sources: for the book you are currently reading, I have translated any quotations originally in German into English. It should be noted that my positionality (cf. Castelló 2022) as an Austrian is apparent in this book, since I often discuss the legal and moral situation in this country. Moreover, it is evident that I have been socialized in a Western capitalist democracy, which heavily influences how I perceive and understand the world. Before we find ourselves in medias res, I would like to tell you a little about my own personal story, since this will help to contextualize the discussions offered in this book.

I grew up in a village in the countryside in Upper Austria. Nonhuman animals (especially cats and horses) have played an important part in my life from my early childhood onward. I became a vegetarian when I was 12 years old, and had a vegan phase of a few months when I was 16, after which I switched back to vegetarianism. Ten years passed before I went vegan again, since which time I have remained a vegan. In the rural environment in which I lived back then, my ethical perspective concerning animal ethics always marked me out as something of an outsider and my

D. Schneeberger, *Envisioning a Better Future for Nonhuman Animals*, The Palgrave Macmillan Animal Ethics Series, https://doi.org/10.1007/978-3-031-75332-9_1

family, who enjoyed eating meat and animal products, found themselves perplexed by my choices. Back then, restaurants didn't even offer a vegetarian dish on their menu as a matter of course, let alone a vegan one. Thankfully, menus have changed as the years have passed, even in Upper Austria. Since my parents found themselves able to support my education, I attended a private high school. After high school, I wanted to attend university and study the subjects which had interested me the most in high school (with philosophy, psychology, English, and German on my list). I opted for a course in English and philosophy, with no idea of whether I would be successful at university or not. I was a first-generation student; no one in my extended family had studied before me, but my parents made it possible for me to study from 2007 to 2016. I completed a Master's in Philosophy and one in English and American Studies, and wrote the philosophical PhD thesis on which this book is based. Back then, I did not hold a position at university and you might say that I was an "academic amateur," enjoying the great privilege of being able to work philosophically, even though I was not paid for this work. In 2019, I obtained a position as a university assistant at the Vienna University of Economics and Business, which enabled me to become a professional scholar in the sense that I could now make a living by doing research. This position provided me with important resources such as further academic education, travel funding, mentors and peers, allowing me to grow significantly as a scholar and to improve my thinking and writing. I wrote a second PhD, which I completed in 2024, this time in the field of economic and social sciences, on the subject of animal organization studies, with my topic being nonhuman animals in organizations. When I had completed my philosophical PhD thesis back in 2016, it was published only in German, in the university library in Salzburg. So I am thrilled that this book—which had its genesis in that thesis—has now "found a home" and can, I hope, open up new perspectives for a broader audience of humans interested in nonhuman animal rights.

1.2 How this Book Contributes to the Literature

This book presents an exercise in anti-speciesist thinking and envisioning. It aims to provide a counterweight to the speciesist world we currently live in. By reducing our speciesism—that is, discrimination against beings due to their species membership (Horta 2010)—we can make this world a better place for nonhuman individuals who are deeply affected by human actions and thinking.

First of all, this book offers general preliminary reflections about possible future universal nonhuman animal rights declarations. It addresses a gap in the literature on nonhuman animal rights by providing a framework that articulates the rights of nonhuman animals in a manner analogous to human rights declarations. When deliberating on possible strategies towards a desired state of the world, it is helpful to understand the status quo. This book therefore also offers descriptive insights into the status quo of how nonhuman animals are treated currently by humans and sheds light on the conditions currently experienced by nonhuman animals used in the animal industry. Moreover, insights into the current legal situation are provided. Manifold jurisdictions exist around the globe; I will primarily focus my discussions about the legal status quo on Austria, the country in which I was born and currently live. Furthermore, when discussing animal rights, some knowledge of the nature of certain beings is essential; this book therefore also discusses biological and ethological knowledge and discusses what follows ethically from this knowledge. I also present insights into the literature on animal rights and animal ethics,[1] and have also included the thoughts of several animal ethicists situated in the German-speaking world. By including these reflections here, I hope that they might reach a wider international audience. In addition, I also present my own ethical normative argumentation and reasoning. Due to the expansiveness of the interdisciplinary discourse about nonhuman animal rights, this book takes a meta perspective on a discourse which is customarily nuanced and detailed, providing a "bird's-eye view" of academic discussions about nonhuman animal rights. That said, at certain points, this book will take a closer look at selected detailed questions. This book then presents an effort to contribute to furthering our thinking about nonhuman animal rights, which it does by making use of diverse aspects of academic work. At times, it is more of a literature review of animal ethics; at times, my own normative argumentation and reasoning is foregrounded, whilst at others, I hone in on the legal situation, or on ethological and biological knowledge. I am convinced that all of these aspects can be instructive: the endeavour that we are embarking on is, after all, highly interdisciplinary.

[1] Since I find the exact linguistic form scholars use to convey their thoughts to be important, I was inspired by Castelló's (2022, 13) notion of "giving authors space to speak for themselves"; I have therefore often included direct quotations in my reflections.

When it comes to nonhuman animal rights, one might ask: Which groups of nonhuman animals should have which rights under which circumstances? And how can we grant those rights? There are thus two steps: deliberating on and deciding what should be the goal, and then formulating a plan to attain that goal. I will try to provide some groundwork that might prove useful on the path towards answering these questions. Just addressing the first question is highly complex: currently, there exist about 8.7 million species of animals, plants, and fungi on this planet, all interwoven in a complex web of existence (Mora et al. 2011). Unsurprisingly, this book does not seek to provide an answer to these all-encompassing questions, but I will focus on providing answers to many questions that come up during the discourse such as: How could we ground nonhuman animal rights theoretically? Which rights could be relevant for many species of animals? Which steps could we take in order to build a world in which we could protect these rights?

1.3 Who Might Find Value in this Book and How to Read this Book

This book might be useful to all those who are interested in the topic as well as providing a good starting point for all those new to the discussion. It might thus be helpful for interested students in the academic field and those with a more general interest alike. Critical animal scholars, ethicists, lawyers, and legal scholars working in the field of nonhuman animal rights might also find this book invaluable. Whilst those new to the topic might appreciate the overview and synthesis of different approaches this book provides, others might appreciate the chapters in which this book goes beyond comparing and contrasting different approaches in the literature and offers critical observations that might inspire new ways of animal ethical thought.

With regard to how to read this book I suggest that you, as the reader, might first wish to look at the subsection titles and decide what is of interest to you, personally. At the end of each section, a few paragraphs provide a summary of the next chapter. After reading this summary, you might decide either to read or to skip a section. For some readers, the normative parts of the book will prove the most interesting, whilst others might be keen to explore empirical realities or biological knowledge. In order to avoid being "armchair philosophers," we need to seek (embodied) insights into the real world and knowledge from other disciplines, and let them inform our views.

Certain topics are mentioned repeatedly in the book, since they are relevant across several domains. For example, the castration of domestic and companion nonhuman animals is relevant both in terms of the right to physical integrity, and in terms of a right to sexuality and procreation. These overlaps mean that certain aspects appear in the discussion of more than one right. Readers might want to consult the index at the back of the book to look for terms and topics of interest and where to find them in the book.

1.4 How this Book Is Structured

Before moving on to the discussion of individual animal rights for nonhuman animals in this book, a number of chapters offer reflections that will provide a basis for the normative argumentation in the main part of the book. First, in the second chapter of this book, which follows the one you are reading now, there are remarks on language in animal ethics as well as metaethical considerations, and a discussion of the ethical principles which form the basis for the normative animal ethics argumentation presented here. In the third chapter, I provide a brief overview of the historical development of animal ethics. I then discuss the role of technology when it comes to making nonhuman animal rights a reality in the fourth chapter. In the fourth chapter, I also shed light on what I believe could be four pillars of nonhuman animal rights. After discussing this, I proceed with some reflections on moral outrage and the "Holocaust analogy." In the fifth chapter, I provide some considerations as to why we need universal declarations of nonhuman animal rights, and how we could develop them. This is followed by a brief sketch of the history of the development of human rights. I then sketch possible parallels between the Universal Convention on the Rights of Persons with Disabilities[2] and potential

[2] Possible future universal animal rights for certain nonhuman animals could be seen as a moral progression that follows on from universal human rights, universal children's rights and universal rights for disabled humans, because they would protect a further group of (certain) vulnerable beings who have interests that need to be considered and protected. The reflections on possible future universal declarations of nonhuman animal rights presented here will, where appropriate, draw parallels between universal children's rights, universal rights for disabled humans, and possible future universal nonhuman animal rights. The

future nonhuman animal rights declarations before I present a concise discussion of the argument of species overlap. Three concepts (needs, interests, capabilities), which are of central importance to this book, are then considered in the sixth chapter. This includes a discussion of the possibility and ethical assessment of interventions in needs, abilities and interests and thus in the nature of nonhuman animals. In the seventh chapter, I provide some reflections on the idea of a just multispecies society.

The main part of this book consists of Chaps. 8–17. Here, I first offer an overview of 12 Articles which could be a part of possible future universal animal rights declarations and then discuss them in more detail. I prevail upon the patience of the reader who may make their way through those preliminary reflections prior to discussing those Articles in individual detail; note that I offer more expansive discussions of some rights than others. These rights include a right to life; a right to freedom; a right to safety/physical integrity/freedom of torture; a right to recognition as a person/(legal) subject before the law; a right to a human guardian, a legal remedy, and an indirect right of legal action; a right to love, sexuality, procreation, pair-building, family, and friendships; a right to property; a right to political consideration and participation; a right to education; a right to work; and finally a right to an adequate standard of living and certain social benefits. Following the discussion of the individual rights, the limitations of this book and avenues for future research are illuminated in the conclusion.

> The day may come when the rest of the animal creation may acquire those rights which never could have been taken from them but by the hand of tyranny.
>
> Jeremy Bentham, *Introduction to the Principles of Morals and Legislation* (1780), Chapter 17, No. 122.

Universal Convention on the Rights of Persons with Disabilities in particular offers a variety of ethical insights that may also help the progress of animal ethics. Furthermore, in the discussion of a nonhuman animal right to love, sexuality, procreation, pair-bonding and family, a section is included explaining the human right to these fundamentals that disabled people have (as stated in the universal Convention on the Rights of Persons with Disabilities). Considering the rights of humans with disabilities might help us to reflect on how we might protect the rights of nonhuman animals in this area.

To us it seems incredible that the Greek philosophers should have scanned so deeply into right and wrong and yet never noticed the immorality of slavery. Perhaps 3000 years from now it will seem equally incredible that we do no notice the immorality of our own oppression of animals.

Brigid Brophy, cited in Richard D. Ryder, *Animal Revolution: Changing Attitudes Towards Speciesism* (2000, pp. 5–6)

We are doing to sentient beings exactly what we would condemn as the worst atrocities if they were done to us. Why do we judge the situation so differently when we are not the (nonhuman) victims but the (human) perpetrators?

From this book.

References

Castelló, Pablo P. "The Language of Zoodemocracy: Contesting Human Sovereignty Over Animals." PhD thesis. Royal Holloway, University of London, 2022.

Horta, Oscar. "What is Speciesism?" *Journal of Agricultural and Environmental Ethics* 23 (2010): 243–266. DOI: https://doi.org/10.1007/s10806-009-9205-2.

Mora, Camilo, Derek P. Tittensor, Sina Adl, Alastair G.B. Simpson, and Boris Worm. "How Many Species Are There on Earth and in the Ocean?" *PLoS Biology* 9, no. 8 (2011): e1001127. DOI: https://doi.org/10.1371/journal.pbio.1001127.

Ryder, Richard D. *Animal Revolution: Changing Attitudes Towards Speciesism.* Oxford: Berg, 2000.

CHAPTER 2

Preliminary Metareflections

I would like to begin the book proper with a few preliminary metareflections in terms of anti-speciesist language, metaethics, and ethical principles. Language plays a pivotal role in shaping our perceptions and actions towards nonhuman animals, as highlighted by Pablo Castelló (2022). In this section of the book, I engage with various animal ethicists' linguistic choices, illustrating progressive shifts towards less speciesist language. These linguistic adjustments aim to challenge anthropocentric views as well as promote greater respect for nonhuman beings. In fact, the discussion extends to the term "animal" itself, critiquing its exclusion of humans and highlighting the problematic human-animal divide ingrained in Western thought. I advocate for a broader, more inclusive understanding that acknowledges humans as animals, emphasizing the interconnectedness of human and nonhuman animal rights. Regarding metaethics, I take a cognitivist, realist, naturalist stance, asserting the existence of ethical truths rooted in natural properties. I discuss ethical principles such as beneficence, non-maleficence, and anti-speciesism before finally discussing the intuitive appeal of sentientism and pathocentrism.

D. Schneeberger, *Envisioning a Better Future for Nonhuman Animals*, The Palgrave Macmillan Animal Ethics Series,
https://doi.org/10.1007/978-3-031-75332-9_2

2.1 On the Language Used in this Book and on Language and Animal Ethics in General

Since our thinking creates our language, our language guides our thinking and our thinking justifies our actions, we ought to be highly mindful of our linguistic choices. As Pablo Castelló (2022, 1) has shown, "language plays a crucial role in producing violence against animals: it determines what we decide; how we feel, think and act; and structures our legal and political institutions." Castelló (2022, 9) continues: "[W]e humans are born into language, and live in and through it." Quoting Austrian philosopher Ludwig Wittgenstein, Castelló (ibid.) tells us that "the limits of our imagination are circumscribed by our language," further elaborating (2022, 74) that: "[O]ur gaze, and our conceptual senses more generally, always sees within Western language. This conceptual, ontological, and historically situated gaze sees nonhuman animals as mere objects to be consumed, as subhuman things to be observed, rather than subjects who speak, feel, and respond." In the Western world, we currently find ourselves "unavoidably thrown into a violent world and language" (ibid.). "The conceptual force of Western language shapes our existence, bodily lives, and "instinctive and intuitive knowledge" (Castelló 2022, 79). Castelló (ibid., 78) thus speaks of a "world-forming force of Western language." "[O]ur embodied lives, and hence our practices in themselves, are co-opted by Western language" (ibid., 80).

It is interesting to observe how language has developed and changed in the animal ethics discourse. Frequently, individual ethical positions can be inferred from the linguistic formulations animal ethicists use. It can, for example, be a sign of progress to refer to the sex of a being using gendered pronouns, as Christine Korsgaard does in the example below. The idea is to no longer think of nonhuman animals as things or objects, which can be the effect of using the neutral pronouns "it"/"its," but to consciously use pronouns that indicate the animal's biological sex. When Christine Korsgaard speaks of a nonhuman animal individual, for example, she chooses words which indicate the gender of a being; cf. e.g. Korsgaard (2011, 100; my emphasis): "A tiger who stands downwind of *her* intended prey is not merely aware of *her* prey—*she* is also locating *herself* with respect to *her* prey in physical space."

The terms used by Austrian animal ethicist and activist Martin Balluch, who, when writing about nonhuman primates, uses the term "man" and "child(ren)," might be initially somewhat confusing, since we are automatically conditioned to think of humans when reading those terms. This linguistic decision may be unusual and may take some getting used to, but at the same time it is a step towards less speciesism and greater respect for nonhuman primates. Here is an example from a book by Martin Balluch (2014, 110; my emphasis):

> The cases of adoption are particularly impressive. Fothergill and Linfield (2012) documented a case where a mother was killed by a leopard. As a result, an adult man took the child, cared for it as it would have been cared for by the mother, and carried it around on his back. This is the only way many orphans can survive. [Christophe] Boesch [2012] mentions a *man* who had adopted three children at the same time. He wanted to take one of the children with him as soon as the mother died, but it stayed with the dead mother. The *man* then came back the next day especially for this purpose, and this time the child willingly let itself be helped and adopted.

By contrast, in the 1994 German translation of Peter Singer's *Practical Ethics* (1979), *Praktische Ethik*, the terms "male" and "female" are encountered respectively; cf. e.g. Singer (1994, 153f; my emphasis) on primates: "Once a young *male* attracted a *female* in just this way when the dominant *male* passed by."

In general, the category "nonhuman animal" is very broad: it includes incredibly diverse forms of life—from sponges to killer whales. At some point in our thinking it might make sense to use this broad category, and at other times we need to be more concrete and specific when referring to a subgroup of this broad category. By contrasting human animals with nonhuman animals, we divide the animal world into two groups, creating an anthropocentric dichotomy. While this can be open to criticism, and in the knowledge that we need to continually advance our thinking and our language, I consciously opt to use this dichotomy in this book.

That the term "animal" is commonly used in a way that does not include the species Homo sapiens is an increasingly antiquated linguistic convention that can be judged as speciesist and biologically inaccurate. Castelló (2022, 156) states that the fact "that we humans are animals has remained concealed for millennia." The human-animal divide is a "central dualism structuring Western thought," dating back to Aristotelian notions of a natural hierarchy and the Christian belief that humans are the crown of the divine creation (Kim 2011, 320). Castelló (2022, 38) emphasizes the epistemic violence inherent in this divide. Referring to Dinesh Wadiwel's work, Castelló (2022, 11) provides further insights:

> Wadiwel argues that Westerners have traditionally adopted (and continue to adopt) a position of human sovereignty, or dominion, over animals at the epistemological level and that it is from that vantage point that we have constructed ethics [...]. The human-animal dichotomy illustrates perfectly what Wadiwel means by epistemic sovereignty. Westerners' thinking usually operates within an epistemology in which the paradigmatic human person, understood as a white and able-bodied man who is rational, independent, and "civil," is at the centre. From this epistemic vantage point, Westerners then decide who can be violated, killed, and excluded from the political community and from moral considerability. My contention is that adopting a position of epistemic sovereignty is unjust and produces unimaginable violence against those who deviate from the human.

In contrast to most established linguistic conventions, when the term "animal" is used in this book, it includes humans; if I refer to fauna excluding humans, I use the term "nonhuman animals." This can on occasion seem linguistically cumbersome, but it affords the advantage of opening new space for anti-speciesist reflection. Various contemporary scholars in animal ethics, however, use the term "animals" to refer exclusively to nonhuman animals for pragmatic, stylistic reasons. That said, to remind us that humans are also animals and that human rights in this sense are a *special* kind of animal rights (Balluch 2005), it can be useful to contrast human animal rights and nonhuman animal rights.

Reflecting on terminology, Stucki (2023, 10) states that "human rights are animal rights and animal rights are human rights." Whilst the first part of the sentence is relatively straightforward, it is not immediately apparent why all animal rights might be supposed to be human rights. Categorically speaking, one could argue that equine rights, or feline rights, which are "animal rights," are *not* human rights. Stucki (2023, 94) elaborates, however, and speaks of an "evolving terminology of 'human rights': [...] animal rights in this broad sense are post-human rights." Even though framing nonhuman animal rights as "new human rights" may be anthropocentric (Bernet Kempers 2023), teleporting ourselves into a possible future, and speaking historically, we could argue that human rights were the first animal rights, and thus one could say all animal rights that came after human rights were, in a sense, (post-) "human rights." Stucki's reflections are embedded in a "one rights approach," inspired by the discussions of "one health" or "one welfare" when it comes to the human and nonhuman fauna on this planet (Stucki 2023). Since—as I argue—we are still in the phase of laying the intellectual groundwork when it comes to nonhuman animal rights, a diversity of linguistic and conceptual options is to be expected, and offers value in terms of the ways in which the terminology in our discussions could develop. However, for the reflections presented in this book, Balluch's (2005) conceptualization as animal rights as the broader category which consist of rights specific to diverse species (human, feline, equine, etc.) seems to be the most useful.

Analogous to the discussion around using the term "nonhuman animals," the scientific field currently referred to as "animal ethics" could be referred to in a less speciesist and more biologically correct way as "nonhuman animal ethics." In this sense, any ethics that addresses ethically good and bad actions concerning animals would then also be called "animal ethics." Such a linguistic change would correspond to a consistent anti-speciesism. However, for pragmatic reasons, and due to the fact that linguistic innovations must grow organically, the traditional definition of "animal ethics" is retained here. (By this token, terms such as "animal industry" or "animal welfare" are also used in this book.)

Another way of referring to nonhuman animals is to speak of "the *other* animals." This linguistic construction refers to the biological fact that humans are one animal species among *other* species. Christine Korsgaard (2011, 111) opts for this type of designation:

> Even in the scientific literature, the term "human" is often used contrastively with "ape" or "primate" or "animal". Of course one may argue that such locutions are mere conveniences—it is so tedious always to say "the other animals" or "the other primates". But when thinking about the ethical treatment of animals, it is useful to remind ourselves that we are among them, so I will say "the other animals" when that is what I mean.

"The other animals" potentially encompasses other human and nonhuman animals; when using it, we could thus be referring to other human animals and/or nonhuman animals. It is broader than the term "nonhuman animals," which excludes human animals. Although the term "nonhuman animals" has the advantage that it usually describes the intended content sufficiently precisely, as mentioned above, it reflects an anthropocentric view since it suggests a dichotomy: humans and *non*human animals. Comparable to this anthropocentric classification is an androcentric pair of terms, namely "male humans" and "non-male humans," or the carrot-centric example "carrots" and "non-carrot vegetables," which reveals a "carrot-centric" worldview (cf. Kompatscher 2015).

The fact that our language is (still) speciesist is visible in the different designations for the same things, actions and conditions in humans and nonhuman animals. Humans *eat, nonhuman* animals *feed*, humans have a *mouth*, nonhuman animals have a *muzzle.* Nonhuman animals are *slaughtered* or *killed*, humans are *murdered.* Our idioms also tell stories about our violent relationship with nonhuman animals. The animal rights organization People for the Ethical Treatment of Animals (PETA) argues for the avoidance of violent language by suggesting using nonviolent idioms, such as feeding two birds with one scone,[1] having bigger fish to free, and not putting all your berries in one basket (PETA n.d.; cf. also Fomenko 2023).

Reducing speciesism in our language can also partly be achieved by reducing the objectification of nonhuman animal individuals by choosing plural forms that point out that we are talking about individuals. Some animal ethicists thus use, for example, "sheeps" instead of "sheep" and "fishes" instead of "fish" when referring to a multitude of beings belonging to these species.[2]

When it comes to "wild animals," in line with discussions about wild animal suffering, I refer to them as "nonhuman animals living in the wild" in this book. The term "wild animals" might be associated with naturalistic, hands-off intuitions which might cause us to care less about their suffering and wellbeing. "Nonhuman animals living in the wild" does not have these associations. Similarly, Charlotte Blattner and colleagues (Blattner et al. 2020, 18) point out that "[u]sing 'farmed' rather than 'farm animals' is [...] one small way to disrupt ingrained habits of thought that assume these animals are born with a functional purpose, and are indeed incapable of living any other kind of life."

In view of the above linguistic considerations, in the following considerations of nonhuman animal rights, linguistic formulations need to be chosen that strike a balance between not being as speciesist as established norms, and not departing from those norms linguistically to such a degree

[1] I might add that baked goods are normally not conducive to a bird's health and might cause digestive issues: try sunflower seeds instead.

[2] Cf. Castelló (2022, 250): "I have emphasised that no one decides to be born into a Western conceptuality structured by all the previous conceptual forces. When one is born and her/his subjectivity is structured by, for instance, the force of the animal and one then goes to fish for fun with one's family as a child, one's subjectivity is being formed not to see fishes—this grammar (fishes) is, incidentally, by and large absent in animal literature!—but rather the fish, the animal."

that it impairs ease of reading and understanding of what is being said. In the next section, I will elaborate on my metaethical positioning and the ethical principles that have influenced my thinking. As I explain below, my personal stance is cognitivist, realist, and naturalist. I believe that ethical truth exists, that we can know something about it, and that what we ought to do can be determined with respect to natural properties.

2.2 Metaethics

Is there such a thing as ethical truth? And if there is, how can we know what action is right or wrong? Ethicists deal with entities that are not as straightforward as the entities that other scholars might deal with. We need to reflect on the ontology of the matters we discuss, and debate if and how we can know something about them (the epistemology). The importance of making our epistemological and ontological standpoint clear is not limited to descriptive discourse, but also pertains to the normative domain. The main distinctions that can be made in the field of metaethics are the ontological distinction between realism and anti-realism—that is, whether the entities we talk about are real (realism) or not (anti-realism)—and the epistemic distinction between cognitivism and non-cognitivism—that is, whether we are able to know something about them (cognitivism) or not (non-cognitivism). Although non-cognitivism is anti-realist (van Roojen 2023), cognitivism does not automatically imply realism. According to an antirealist cognitivist position, we can acquire knowledge about ethical principles and those principles can be true or false,[3] but they do not exist independently from human agents. Realist cognitivist positions, on the other hand, hold that objective moral facts exist and that we can know something about them. Naturalism, for example, holds that moral facts are parts of nature (ibid.).

When it comes to describing metaethical positions, next to the ontological (realism vs. anti-realism) and the epistemic dimension (cognitivism vs. non-cognitivism), we might then moreover add a third dimension, which can be located in the branch of philosophy of language, the semantic dimension. In this third dimension, we are interested in the meaning and function of ethical utterances. The statement "X is (ethically) good" can, on a linguistic level, be understood as a descriptive utterance stating a

[3] By engaging in a process of idealized deliberation (cf. Habermasian discourse ethics), for example.

fact about an objective entity, or as a prescriptive statement, implying "Do X!" The position that considers seemingly descriptive statements such as "X is good" as prescriptive statements is called metaethical prescriptivism (van Roojen 2023). Metaethical prescriptivists consider "X is good" as implying "Do X!" (ibid.). Metaethical emotivism is antirealist, non-cognitivist, and semantically understands the function of statements such as "X is good" as expressing emotions (ibid.).

In order to think and argue precisely, we need to know whether someone engaging in normative discourse is making a statement that is, for example, metaethically equivalent with how they are feeling today (cf. emotivism[4]) or if this statement is meant as an objective (and appellative) statement. We need to know about these metaethical dimensions in order to be able to decide on an appropriate reaction to this statement. Imagine that one person is arguing that "X is ethically good" and you disagree. It is important to understand whether this person has made a normative, appellative utterance, urging us to do X, or whether this person has merely informed us about an "emotion" they experience. In the case of the latter, it makes little sense to start disagreeing with this person about how they feel, since this person has not made a statement about what we ought to do at all. In the following section, I argue that the pain of sentient beings matters ethically (cf. sentientism, Rodogno 2010) and that pathocentrism does have intuitive appeal for me and influences my reflections. In the next section, I discuss ethical principles which are central in my argumentation (the principle of beneficence, the principle of non-maleficence, and anti-speciesism).

2.3 Ethical Principles

Ethical "truths" and insights can neither be mathematically proven nor inductively replicated again and again in scientific experiments (such as the laws of nature can). Nevertheless, there are widely shared ethical convictions upon which ethical theories are (or can be) based. Human morality is, after all, "grounded in pre-reflective moral sentiments and pro-social impulses that we share with many animals" (Donaldson and Kymlicka

[4] "Emotivists think moral terms in grammatically assertive utterances function primarily to express emotion" (van Roojen 2023). Those utterances can thus not be true or false. However, the person expressing their "emotion" in an appellative fashion might thereby wish to influence the other's "emotion(s)" (ibid.).

2014, 26). There are (animal) ethical (basic) intuitions[5] that guide us in everyday life: these intuitions have become principles that are widely shared and agreed upon across many cultures and religions. There may be exceptional cases where these principles are not unconditionally valid; however, by and large they are endorsed.

One example of such an intuitive ethical principle is, ceteris paribus (that is, all else being equal): "It is good to help someone in need and bad to harm someone" (cf. the principle of beneficence and the principle of non-maleficence in biomedical/clinical ethics (Beauchamp and Childress 2013; Varkey 2021)) or as Arthur Schopenhauer put it (Schweppenhäuser 2005, 174): "Hurt no one, rather help everyone as far as you can." There is no reason why these principles should stop at the species border. It is good to treat a dog in such a way that supports the dog's wellbeing, and bad to maltreat a dog. Ethical intuitions might tell us that it is good to rush to the aid of an injured person who has had an accident (generally speaking; exceptions and special constellations are excluded here for the time being), and bad to abandon someone who needs help, if we are in a position to provide it without too much of a cost to ourselves. When deciding how to act, it is good to consider the wellbeing and suffering of others. It is generally morally better to include the interests of other sentient beings in one's decisions to act, rather than to selfishly pursue and enforce one's own interests. Inflicting suffering, harm and pain (senselessly) is ethically worse than promoting health, wellbeing, contentment and happiness. We have intuitions that tell us that it is generally better to save as many humans as possible from a burning house, rather than setting fire to yet another house instead, thus causing even more suffering, pain, and ruin. It is better to resolve (international) conflicts peacefully, where possible, than it is to increase suffering, distress, death and damage through warlike violence.

In summary, firstly (and still broadly speaking), widely shared basic moral intuitions tell us that those acts which promote health, flourishing and wellbeing, rather than those acts which increase pain, suffering and death, are morally good. The argumentation presented in this book also

[5] Cf. on the concept of intuition Franklin's (2005, 24) distinction between two types of intuitions: "There is a great difference between immediate intuitive response [...] and reflective intuition. The first is prereflective and can vary widely in different individuals. The second is not an immediate response to an individual or individual situation. It is the outcome of careful reflection on a principle which prereflective intuition recommends."

includes various naturalistic influences, especially those of neo-Aristotelian thinkers like Martha Nussbaum and Bernard Rollin. Not least through G.E. Moore's remarks on naturalistic fallacies, ethical naturalism is met with scepticism (cf. e.g. Kallhoff 2010, 20). On closer examination of what is or can be meant by naturalistic normative ethical arguments, it becomes clear that the term "ethical naturalism" can be used to describe a whole spectrum of positions (Kallhoff 2010, 19). Kallhoff (ibid., 25) explains: "An ethical naturalist does not claim that something is good because and insofar as it is natural. Rather, an ethical naturalist claims to have a theory that shows that right action can be determined with respect to natural properties." Thus, few ethicists who assume that the "right action" can be determined (or that there are at least better and worse alternative actions) would dispute this almost trivial claim, which is also central in my own argumentation: how we ought to treat sentient beings can be determined (at least to a considerable extent[6]) with regard to their sentience, their ability to experience pain and wellbeing. In the sense of Bernard E. Rollin's (Rollin 2011, 51) statement, "[S]atisfaction of telos results in happiness for an animal," it is important to highlight the connection between the natural capacities of a living being, i.e. its needs, interests, and capabilities, and the ethically good or preferable treatment of this being.

Pathocentrism has intuitive appeal. It is often precisely the characteristic of sentience—the ability to feel pain—that we humans share with many other animal species. This ability is ultimately based on natural biological facts, and because of this basis, it can be thought of as naturalistic in a certain sense. The basis of pathocentric arguments is, on the one hand, that sentient beings can naturally feel pain and, on the other hand, that it is ethically good not to cause pain (and, on the contrary, to promote wellbeing). I argue that the pain of sentient beings matters ethically.

The fact that pathocentric arguments have this basis "in the natural" does not mean that these arguments are weak, flawed or even invalid because of this biological basis, which refers to the nature of beings and the (ethically relevant) natural quality of their sensations. In this sense, a

[6] Although sentience might serve us in our debate about moral status, I agree with Castelló (2022, 164) when he makes the case that "[...] [v]alue flows from honeybees even if they are not sentient in the terms that authors like Singer think of sentience. For this reason, sentience, at least understood as feeling pain and pleasure in the exclusively physiological-physical sense, is not a necessary condition to be worthy of moral considerability."

certain naturalism can be helpful, convincing, and easy to understand in ethical-normative argumentation. I am in agreement with Peter Singer, who states that equal interests are to be considered equally, ceteris paribus, regardless of who the bearer of those interests is (cf. Peter Singer's principle of equal consideration of interests: "The essence of the principle of equal consideration of interests is that in our moral deliberations we give equal weight to the similar interests of all those who are affected by our actions" (Singer 1994, 39)).

Discrimination against individuals or groups, i.e. disadvantaging or degrading them, is the cause of numerous ethical shortcomings. Discrimination against *humans* on the basis of diverse but morally irrelevant characteristics is now frequently expressly prohibited in advanced Western societies (cf. Article 2 of the Universal Declaration of Human Rights (UDHR): "Everyone is entitled to all the rights and freedoms set forth in this Declaration, without distinction of any kind, such as race, colour, sex, language, religion, political or other opinion, national or social origin, property, birth or other status"). Nevertheless, racist, xenophobic or sexist tendencies, although commonly recognised as morally bad, persist and are hard to tackle; in particular, apartheid and slavery are historical[7] phenomena that were strikingly racist. People who are categorized as "non-white" are frequently still discriminated against today, both in the US and Europe, as well as elsewhere. In any number of countries, discrimination is frequently based on nationality or ethnicity (nationalism, xenophobia, xenophobia). Younger and older humans sometimes encounter prejudice because of their age (ageism). Additionally, unequal treatment of humans on the basis of their religion, their sexual orientation, their individual constitution (e.g. being overweight, or because of specific facets of appearance or impairment) is considered morally reprehensible.

Peter Singer's principle of the equal consideration of interests (1994) states that it is wrong not to give equal weight to the interests of nonhuman animals which are sufficiently similar to corresponding human interests. It can be speciesist to give greater weight to the interests of human individuals than to the interests of nonhuman living beings. Speciesist

[7] Slavery, unfortunately, is still an acute societal issue in many countries (cf. also Blattner and Bossert 2022, 25). In 2016, "24.9 million people were in forced labour. That is, they were being forced to work under threat or coercion as domestic workers, on construction sites, in clandestine factories, on farms and fishing boats, in other sectors [...]. Forced labourers produced some of the food we eat and the clothes we wear [...]" (International Labor Office 2017).

views are often argued for on the grounds that humans supposedly embody greater abilities and more potential than do nonhuman animals, and that a human life is therefore worth more than a nonhuman life. This shows that speciesism is entangled with ablesism (Taylor 2017). Diverse forms of domination—such as racism, speciesism, and sexism—are co-constituted and mutually sustaining (cf. Ko and Ko 2017). Since unjust discrimination should be ethically opposed, and speciesism is unjust discrimination, anti-speciesism should be incorporated into our value system, just as anti-racism[8] and anti-sexism are.

Speciesism, however, might have a relatively strong intuitive basis due to the ascription of different values to lives of beings due to their species membership. What kind of value do we ascribe to different beings and their individual lives, in thought experiments, for example?[9] If you had to decide, for example, to save five humans or five fruit flies from a burning house, which choice would you make? It should perhaps be borne in mind that fruit flies have nociceptors, which could point towards a certain sensitivity to pain (cf. e.g. Tracey et al. 2003; Knutsson 2016). Is it speciesist to choose humans? We might argue that we know more about what it's like be a human: we know about the "inner world" of humans, while we know very little about what it's like to be an insect (Earp 2017; Balcombe 2021; Abbate 2023). Since we "can never fully know, [...] we ought to be more than careful about what we judge another's life to be" (Castelló 2022, 204f). It could be intuitive speciesism that guides our decision to save the humans, but also our tendency to treat nonhuman animals who are small worse, since we value them less (cf. "sizeism" (Morton 1998; MacClellan 2013)). Is it speciesist to administer an anthelmintic to a child with a severe case of worms, thereby promoting its—one—life, but forfeiting that of the (many) thriving worms? (In the case of the child suffering from worms, a right to the life of the worms could, however, arguably be nullified by the ethically justified (reasonable) self-defence on the part of the child). Absolute value-egalitarianism when it comes to the value of a sentient life might be overwhelming and may lack intuitive backing, even if one is fundamentally in favour of a right to life for (very) many species. I have merely touched upon the debate about the value of the life of beings here; it will be taken up in detail later on in this book.

[8] Cf. e.g. Ko (2017).

[9] Thought experiments might help us to envision better futures and grow morally (Cooke 2017).

In the next chapter, I first explore the history and modern debates on animal rights, from ancient philosophers to contemporary thinkers like Peter Singer, Tom Regan, and Martha Nussbaum. One key historic debate in animal ethics is between welfarism—which aims to reduce animal suffering within human use—and abolitionism, which advocates for ending all use of nonhuman animals by humans. Peter Singer's "Animal Liberation" and Tom Regan's "The Case for Animal Rights" have been influential in promoting neo-Kantian and abolitionist arguments. Feminist ethicists have criticized traditional rational-focused approaches as patriarchal and speciesist. The discourse has evolved to consider positive rights for nonhuman animals beyond mere protection of life. Scholars like Eva Meijer stress the importance of recognizing the shared planetary existence between human and nonhuman animals. The "political turn" in animal ethics, which has been heralded by works such as *Zoopolis* by Will Kymlicka and Sue Donaldson, proposes a post-exploitation multispecies society, emphasizing political strategies to reshape societal norms.

References

Abbate, Cheryl E. "What it's like, or not like, to Bee." *Between the Species* 26, issue 1 (2023): 1–17.

Balcombe, Jonathan. Super Fly: *The Unexpected Lives of the World's Most Successful Insects.* London: Penguin Books, 2021.

Balluch, Martin. *Die Kontinuität von Bewusstsein: Das naturwissenschaftliche Argument für Tierrechte.* Vienna: Guthmann-Peterson, 2005.

Balluch, Martin. *Der Hund und sein Philosoph: Plädoyer für Autonomie und Tierrechte.* Vienna: Promedia, 2014.

Beauchamp, Tom L., and James F. Childress. *Principles of Biomedical Ethics.* New York: Oxford University Press, 2013.

Bernet Kempers, Eva. "Review: Saskia Stucki, One Rights: Human and Animal Rights in the Anthropocene." *LEOH – Journal of Animal Law, Ethics and One Health* (2023): 24–28. DOI: https://doi.org/10.58590/leoh.2023.002

Blattner, Charlotte E., and Leonie N. Bossert. "Tierarbeit – Animal Labour – Der Versuch einer Einordnung." *TIERethik* 24 (2022): 8–38.

Blattner, Charlotte E., Sue Donaldson, and Ryan Wilcox. "Animal Agency in Community: A Political Multispecies Ethnography of VINE Sanctuary." *Politics and Animals* 6 (2020): 1–22.

Boesch, Christophe. *Wild Cultures: A Comparison between Chimpanzee and Human Cultures.* Cambridge: Cambridge University Press, 2012.

Castelló, Pablo P. "The Language of Zoodemocracy: Contesting Human Sovereignty Over Animals." PhD thesis. Royal Holloway, University of London, 2022.

Cooke, Steve. "Imagined Utopias: Animals Rights and the Moral Imagination." *Journal of Political Philosophy*, 25 (2017): e1–e18. DOI: https://doi.org/10.1111/jopp.12136.

Donaldson, Sue, and Will Kymlicka. "Unruly Beasts: Animal Citizens and the Threat of Tyranny." *Canadian Journal of Political Science / Revue Canadienne de Science Politique* 47, no. 1 (2014): 23–45.

Earp, Brian. "What is it like to be a bee?" *Think* 16, no. 45 (2017): 43–49. DOI: https://doi.org/10.1017/s1477175616000336.

Fomenko, Olena. "Of Mice and Men: PETA's Animal-Friendly Idioms as a Strategy Against Speciesism?" *Acta Linguistica Lithuanica* 88 (2023): 165–180. DOI: https://doi.org/10.35321/all88-08.

Fothergill, Alastair, and Mark Linfield, directors. *Chimpanzee*. Disneynature. 2012.

Franklin, Julian H. *Animal Rights and Moral Philosophy*. New York: Columbia University Press, 2005.

International Labor Office. "Global Estimates of Modern Slavery." Published 2017. Accessed March 23, 2024. https://www.ilo.org/wcmsp5/groups/public/@dgreports/@dcomm/documents/publication/wcms_575540.pdf.

Kallhoff, Angela. *Ethischer Naturalismus nach Aristoteles*. Paderborn: Mentis, 2010.

Kim, Claire Jean. 2011. "Moral Extensionism or Racist Exploitation? The Use of Holocaust and Slavery Analogies in the Animal Liberation Movement." *New Political Science* 33 (3): 311–33. Doi: https://doi.org/10.1080/07393148.2011.592021.

Knutsson, Simon. "Reducing suffering among invertebrates such as insects: Policy paper." *Sentience Politics* 1 (2016): 1–18.

Ko, Aph, and Syl Ko. *Aphro-ism: Essays on Pop Culture, Feminism, and Black Veganism from Two Sisters*. Brooklyn, NY: Lantern Books, 2017.

Ko, Syl. "'By 'Human,' Everybody Just Means 'White.'" In *Aphro-ism: Essays on Pop Culture, Feminism, and Black Veganism from Two Sisters*, edited by Aph Ko, and Syl Ko, 20–27. New York: Lantern Books, 2017.

Kompatscher, Gabriele. Lecture entitled "Human-Animal Studies: New Perspectives of Animal-Human Relationships." University of Salzburg. December 8, 2015.

Korsgaard, Christine M. "Interacting with Animals: A Kantian Account." In *The Oxford Handbook of Animal Ethics*, edited by Tom L. Beauchamp, and R. G. Frey, 91–118. Oxford: Oxford University Press, 2011.

MacClellan, Joel. "Size Matters: Animal Size, Contributory Causation, and Ethical Vegetarianism." *Journal of Animal Ethics*, 3 no. 1 (2013): 57–68. DOI: https://doi.org/10.5406/janimalethics.3.1.0057.

Morton, David B. "Sizeism." In *Encyclopedia of Animal Rights and Animal Welfare*, edited by Marc Bekoff and Carron A. Meaney, 318. Westport, CT: Greenwood Press, 1998.

People for the Ethical Treatment of Animals (PETA). "Bigger Fish to Free: PETA's Complete List of Animal-Friendly Idioms." Last modified n.d.. Accessed February 17, 2024. https://www.peta.org/features/animal-friendly-idioms/.

Rodogno, Raffaele. "Sentientism, wellbeing, and environmentalism." *Journal of Applied Philosophy* 27, no. 1 (2010): 84–99. DOI: https://doi.org/10.1111/j.1468-5930.2009.00475.x.

Rollin, Bernard E. *Putting the Horse before Descartes: My Life's Work on Behalf of Animals*. Philadelphia: Temple University Press, 2011.

Schweppenhäuser, Gerd. *Die Antinomie des Universalismus: Zum moralphilosophischen Diskurs der Moderne*. Würzburg: Königshausen und Neumann, 2005.

Singer, Peter. *Practical Ethics*. Cambridge: Cambridge University Press, 1979.

Singer, Peter. *Praktische Ethik*. Translated by Oscar Bischoff, Jean-Claude Wolf, Dietrich Klose, and Susanne Lenz. Stuttgart: Reclam, 1994.

Stucki, Saskia. *One Rights: Human and Animal Rights in the Anthropocene*. Cham, Switzerland: Springer, 2023.

Taylor, Sunaura. *Beasts of Burden: Animal and Disability Liberation*. New York: The New Press, 2017.

Tracey, Daniel W., Rachel Wilson, Gilles Laurent, and Seymour Benzer. "painless, a Drosophila Gene Essential for Nociception." *Cell* 113, no. 2 (2003): 261–273. DOI: https://doi.org/10.1016/s0092-8674(03)00272-1.

van Roojen, Mark. "Moral Cognitivism vs. Non-Cognitivism." In The Stanford Encyclopedia of Philosophy Archive: Winter 2023 Edition, edited by Edward N. Zalta, and Uri Nodelman. Last modified 2023. https://plato.stanford.edu/archives/win2023/entries/moral-cognitivism/.

Varkey, Basil. "Principles of Clinical Ethics and Their Application to Practice." *Medical Principles and Practice* 30, no. 1 (2021): 17–28. https://doi.org/10.1159/000509119.

CHAPTER 3

Historic and Current Discussions About Nonhuman Animal Rights: Welfarism vs. Abolitionism, Utilitarianism, Deontological Positions, Feminist Animal Ethics, and the Political Turn

In ancient philosophical writings from as early as the fifth century BC onwards, animal ethical considerations can be found, for example, in the writings of Empedocles, Porphyrios, and Pythagoras. Notable philosophical contributions concerning how humans should treat nonhuman animals were later made by Aristotle, Augustine, Thomas Aquinas, Descartes, Kant, Schopenhauer, and Schweitzer, amongst others, with animal ethics later experiencing a renewed upswing in the last few decades of the twentieth century. The reflections in this book are situated within this modern discussion of animal ethics, which is why I would like to first provide an overview of central contemporary thinkers and their positions.

Two strands emerged early on in the modern animal ethics discourse: welfarism and abolitionism. While welfarist positions argue for the avoidance of unnecessary animal suffering when it comes to human use of nonhuman animals, abolitionist positions call for the complete abolition of the use of nonhuman animals by humans. Welfarism is currently closer to the majority opinion of humans in many countries, and in line with nonideal theory approaches in animal ethics. However, welfarism is criticized as being speciesist: "[F]for the sovereign the limits of ethics are open only

D. Schneeberger, *Envisioning a Better Future for Nonhuman Animals*, The Palgrave Macmillan Animal Ethics Series,
https://doi.org/10.1007/978-3-031-75332-9_3

insofar as our sovereign right to dominion and violence is not tempered" (Castelló 2022a, 252). Abolitionists might compare their stance with the abolition of human slavery: they do not want a better treatment of slaves; they want slavery to be abolished, since it is an unjust institution.

It is said that modern animal ethics really began with the utilitarian philosopher Peter Singer who published *Animal Liberation* in 1975. Eight years after Singer's *Animal Liberation*, in 1983, Tom Regan published the book *The Case for Animal Rights*, in which he advocates for a neo-Kantian and abolitionist approach to animal rights that continues to strongly influence animal ethics writings up to this day. In 1995, Gary Francione, an American law scholar, published *Animals, Property, and the Law*. Francione advocates for the strict detachment of nonhuman animals from human society. He holds an extinctionist position, advocating not only for an end of the exploitation of nonhuman animals (abolitionism), but also advocating for the extinction of domesticated nonhuman animals. Francione is also well-known known for his call that nonhuman animals should no longer be categorized as property. In her impactful publications on animal ethics (cf. e.g. Korsgaard 2011), Harvard philosophy professor Christine Korsgaard, similar to Tom Regan, offers a neo-Kantian argumentation. Korsgaard (2011, 110) proposes an animal-ethical norm inspired by the Kantian "categorical imperative," which is similar to the ethical "Golden Rule"—which finds wide cross-cultural approval—whereby nonhuman animals are also included in the moral circle: "Always treat an animal in the way that you can assume he or she would agree to be treated if he or she could."

Martha Nussbaum, Professor of Law and Ethics at the University of Chicago, extended the capability approach, originally conceived by Armatya Sen with regard to humans, to nonhuman animals (Nussbaum 2006). Nussbaum's approach is Neo-Aristotelian, as is the approach of Rosalind Hursthouse (2000, 2011), whose feminist Neo-Aristotelian virtue ethics state that Singer's utilitarianism and Regan's deontological approach actually rely on the two virtues of compassion and respectful love (cf. Köllen and Schneeberger 2023). Virtues provide guidance when choosing how to act. According to Hursthouse (2011), ethics is highly situational and context-specific. Feminist animal ethics scholars have criticized the focus on the rational in the animal ethics discourse as being indicative of the patriarchal powers that dominate academia (Bossert 2018, 117). Targeting speciesism by referring to the centrality of rationality "does not tackle the problem of logocentrism" (Castelló 2022a, 56;

Derrida 2016, 11ff; Sayers 2016). Influential feminist animal ethicists—including Josephine Donovan (2006), Carol Adams (2010), Lori Gruen (Adams and Gruen 2014), and Lisa Kemmerer (2016)—emphasize the importance of relationships, emotions, and contexts when it comes to normative ethics. By pointing to this shortcoming of the traditional modern animal ethics discourse, feminist animal ethicists have added nuance and shifts in perspectives to the animal ethics debate. Castelló (2022a, 56f) explains that

> the gravitational field of the human impels animal ethicists to put the emphasis in [sic!] human-like characteristics (what is like us) and undermine ethico-political dimensions such as: vulnerability (Plumwood 1993; Deckha 2015; Taylor 2017), embodied relationality (Donovan and Adams 2007; Adams and Gruen 2014), dependency (Donaldson and Kymlicka 2011; Taylor 2017), difference (Wolfe 2003, 8; MacKinnon 2007, 264; Castelló 2022b), and relational agency (Donaldson and Kymlicka 2011; Meijer 2019; Blattner et al. 2020): all of which are precisely those characteristics that are not like the human.

With regard to human animal rights, the distinction between negative and positive rights is common: "A positive right entitles the right-holder to have the duty-bearer do some act, while a negative right entitles the right-holder to have the duty-bearer refrain from doing an act" (Wibye 2022). When it comes to nonhuman animal rights, negative rights, such as the right to life, were at the centre of discussions in the Singer/Regan era of modern animal ethics discourse. In the last few years, positive animal rights have become the centre of attention. Dutch philosopher Eva Meijer (2013, 31) emphasizes the importance of positive nonhuman animal rights: "although universal negative rights are important, they are not enough, because [...] humans and animals are cohabitants of the same planet." Hadley (2015, 117) forges a connection between previous generations of animal ethicists and the contemporary one, discussing the "extensionist project" of which this book is a part: "While the task of the first generation was to make the case for the inclusion of animals within the so-called sphere of moral concern, the task of a second generation philosopher, like me, is to take the next step—to examine which rights it is logically possible to extend to animals, assuming that they (animals) do matter." The concept of expanding the moral circle (Singer 2011) is intuitively helpful, even if it is open to criticism for being anthropocentric.

While the discussion of positive rights for nonhuman animals is a comparatively new phenomenon in the animal rights discourse, intuitions supporting positive nonhuman animal rights are already apparent in centuries-old publications. Cato, for example, argued in the last century BC that "giving grace to horses that have become old, and giving care to dogs not only when they are young but also in old age, is the duty of honour of a good man" (Plutarch 2001, 437). The Muslim scholar 'Izz al-Din ibn 'Abd as Salam (Bagader et al. 2008) argued in the thirteenth century AD for a right to fair working conditions and a pension for nonhuman animals used by humans, as well as social security in case of incapacity, and a right to sexuality and procreation: "The rights of livestock and animals upon man [.] are that he spend on them the provision that their kinds require, even if they have aged or sickened such that no benefit comes from them; that he not burden them beyond what they can bear; [.] that he puts their males and females together during their mating seasons [.]."

Although such ancient and prescient intuitions exist, animal scholars addressing positive nonhuman animal rights have, until recently, touched on them only briefly in passing. Looking at the world in which we currently find ourselves, we see a world that still seems some (considerable) distance away from even basic negative nonhuman animal rights. Even if Steve Sapontzis (1987, 83, cited in Donaldson and Kymlicka 2013, 23) opined some time ago that the question of how exploited farmed and laboratory animals "should be treated after they have been released from their present torment [...] is a problem for a much better world than ours," we nevertheless need to consider and test our intuitions about positive rights for nonhuman animals in order to envision a just possible future interspecies society.

Alongside feminist animal ethics, and the discussion of positive nonhuman animal rights, the "political turn" in animal ethics has opened up new ways of reflecting on animal ethics. In 2011, Will Kymlicka and Sue Donaldson published *Zoopolis*, a political theory of what a multispecies society might look like after the abolition of human exploitation of farmed animals. This revolutionary book is a key "turn text": an intellectual contribution to the political turn in animal ethics. Many scholars involved in this political turn in animal ethics are convinced that the cause of nonhuman animal rights needs a political strategy (Milligan 2016). "Turn theorists" have on the one hand engaged in realistic nonideal theory thinking, taking into account the political possibilities in current liberal democracies (ibid.), but on the other hand also developed ideal theory visions, envisioning more just, future multispecies societies (Donaldson and Kymlicka 2011; Cochrane 2013; Castelló 2022b): "It is an important task of political philosophy not just to describe the features of an ideal political society in order to act as a beacon of travel, but also to light the paths we can take to reach it" (Cooke 2017, e1). The objective of this heavily normative interdisciplinary discourse taking place in the animals and politics debate is to "reshape mainstream political processes and to secure the position of animal rights discourse within the regular bounds of political theory" (Milligan 2016, 111). "Turn texts" tend to appeal to liberal values, and put a strong emphasis on animal interests and positive animal rights (i.e. claim rights) (Milligan 2016).

In his 2010 book *An Introduction to Animals and Political Theory*, the British political theorist Alasdair Cochrane (2010, 9) states that the discussion of the fair treatment of nonhuman animals must move from the periphery to the centre of political discourse. In 2013, in *A Theory of Justice for Animals: Animal Rights in a Nonideal World*, British political scholar Robert Garner outlines his ideas for a new political system that entails more justice for nonhuman animals. Some of his ideas and proposals diverge strongly from those put forward in *Zoopolis*. Garner offers a nonideal theory, whereas Donaldson and Kymlicka's theory corresponds to an ideal theory. Garner believes that in designing a political theory that takes better account of the interests of nonhuman animals, one must always keep in mind the status quo of the current world. He regards a theory that is idealistically doctrinaire and does not focus on realistic changes which could actually be achieved to be deficient (ibid.). Garner's reasoning is criticized by Francione as "new welfarism" (Francione 1998; Francione and Garner 2010); whilst Garner demands that no suffering should be inflicted on nonhuman animals—and this demand could in and of itself be construed as very far-reaching—he simultaneously negates that they (should) have a right to life. In terms of the abolitionist ideal that nonhuman animal rights approaches strive to achieve, Garner's approach might not seem sufficiently demanding. His concerns that the abolition of the exploitation of nonhuman animals by humans seems unlikely in the near future, are, however, obviously justified.

When we examine the world around us, we see that despite progress in animal welfare legislation in diverse countries, global meat consumption continues to rise. In the following chapter, I firstly argue that cultivated animal products could offer a promising solution, and a reduction in animal suffering, with developments such as the first cultivated hamburger in

2013 marking significant milestones. Although challenges such as high production costs exist, ongoing research and development efforts aim to address these barriers. Success in adopting cultivated animal products could lead to transformative changes in how we view and treat nonhuman animals, complemented by shifts in societal values towards empathy and ethical consideration of the nonhuman fauna. I argue that technological innovations might disrupt existing systems faster than traditional animal activism and advocacy can.

References

Adams, Carol J. *The Sexual Politics of Meat.* New York: Bloomsbury, 2010.

Adams, Carol J, and Lori Gruen. *Ecofeminism: Feminist Intersections with Other Animals and the Earth.* New York: Bloomsbury, 2014.

Bagader, Abubakr Ahmed, Abullatif Tawfik El-Chirazi El-Sabbagh, Mohamad As-Sayyid Al-Glayand, and Mawil Yousuf Izzi-Deen Samarrai. "Environmental Protection in Islam (Part 6 of 7): Conservation of Basic Natural Elements - Plants and Animals." Last modified February 24, 2008. Accessed April 6, 2024. https://www.islamreligion.com/article/pdf/324.

Blattner, Charlotte E., Sue Donaldson, and Ryan Wilcox. "Animal Agency in Community: A Political Multispecies Ethnography of VINE Sanctuary." *Politics and Animals* 6 (2020): 1–22.

Bossert, Leonie. "Feministische und fürsorgeethische Ansätze." In *Handbuch Tierethik. Grundlagen – Kontexte – Perspektiven*, edited by Johann. S. Ach and Dagmar Borchers, 117–122. Heidelberg: J. B. Metzler, 2018.

Castelló, Pablo P. "The Language of Zoodemocracy: Contesting Human Sovereignty Over Animals." PhD thesis. Royal Holloway, University of London, 2022a.

Castelló, Pablo P. "The erasures of Peter Singer's theory, and the ethical need to consider animals as irreducible others." *Philosophy Today*, 66, no. 1 (2022b): 637–653. DOI: https://doi.org/10.5840/philtoday2022325454.

Cochrane, Alasdair. *An Introduction to Animals and Political Theory.* New York: Palgrave Macmillan, 2010.

Cochrane, Alasdair. "Cosmozoopolis: The Case Against Group-Differentiated Animal Rights." *Law, Ethics and Philosophy* 1 (2013): 127–141.

Cooke, Steve. "Imagined Utopias: Animals Rights and the Moral Imagination." *Journal of Political Philosophy*, 25 (2017): e1–e18. DOI: https://doi.org/10.1111/jopp.12136.

Deckha, Maneesha. "Vulnerability, Equality, and Animals." *Canadian Journal of Women and the Law* 27, no. 1 (2015): 47–70. DOI: https://doi.org/10.3138/cjwl.27.1.47.

Derrida, Jacques. *Of Grammatology*. 4th ed. Translated by Gayatri C. Spivak. Baltimore: John Hopkins University Press, 2016.

Donaldson, Sue, and Will Kymlicka. *Zoopolis: A Political Theory of Animal Rights*. Oxford: Oxford University Press, 2011.

Donaldson, Sue, and Will Kymlicka. *Zoopolis: Eine politische Theorie der Tierrechte*. Translated by Joachim Schulte. Berlin: Suhrkamp, 2013.

Donovan, Josephine. "Feminism and the treatment of animals: From care to dialogue." *Signs: Journal of Women in Culture and Society* 31, no. 2 (2006): 305–329. DOI: https://doi.org/10.1086/491750.

Donovan, Josephine, and Carol J. Adams. *The Feminist Care Tradition in Animal Ethics: A Reader*. New York: Columbia University Press, 2007.

Francione, Gary L. "Animal Rights and New Welfarism." In *Encyclopedia of Animal Rights and Animal Welfare*, edited by Mark Bekoff and Carron A. Meaney, 45. Westport, CT: Greenwood Press, 1998.

Francione, Gary L., and Robert Garner. *The Animal Rights Debate: Abolition or Regulation?* New York: Columbia University Press, 2010.

Hadley, John. *Animal Property Rights: A theory of habitat rights for wild animals*. London: Lexington Books, 2015.

Hursthouse, Rosalind. *Ethics, Humans and Other Animals: An Introduction with Readings*. London: Routledge, 2000.

Hursthouse, Rosalind. "Virtue ethics and the treatment of animals." In *The Oxford Handbook of Animal Ethics*, edited by Tom L. Beauchamp, and Raymond G. Frey, 119–143. Oxford: Oxford University Press, 2011.

Kemmerer, Lisa. "Investigating intersections: Exploring the growth and expansion of anymal liberation." *Green Theory and Praxis* 10, no. 3 (2016): 53–76.

Köllen, Thomas, and Doris Schneeberger. "Avoiding unnecessary suffering: Towards a moral minimum standard for humans' responsibility for animal welfare." *Business Ethics, Environment and Responsibility* 32, no. 4 (2023): 1139–1149. DOI: https://doi.org/10.1111/beer.12565.

Korsgaard, Christine M. "Interacting with Animals: A Kantian Account." In *The Oxford Handbook of Animal Ethics*, edited by Tom L. Beauchamp, and Raymond G. Frey, 91–118. Oxford: Oxford University Press, 2011.

MacKinnon, Catharine A. "Of Mice and Men: A Fragment on Animal Rights." In *The Feminist Care Tradition in Animal Ethics*, edited by Josephine Donovan and Carol J. Adams, 316–332. New York: Columbia University Press, 2007.

Meijer, Eva. "Political Communication with Animals." *Humanimalia* 5, no. 1 (2013): 28–52.

Meijer, Eva. *When Animals Speak: Toward An Interspecies Democracy*. New York: New York University Press, 2019.

Milligan, Tony. "Putting Pluralism First." In *The Political Turn in Animal Ethics*, edited by Siobhan O'Sullivan and Rober Garner, 85–102. Lanham, MD: Rowman & Littlefield International, 2016.

Nussbaum, Martha C. *Frontiers of Justice: Disability, Nationality, Species Membership*. Cambridge, MA: Harvard University Press, 2006.

Plumwood, Val. *Feminism and the Mastery of Nature*. London: Routledge, 1993.

Plumwood, Val. "Human Vulnerability and the Experience of Being Prey." *Quadrant* 39, no. 3 (March 1995): 29–34.

Plutarch. *Fünf Doppelbiographien: Teil 1: Alexandros und Caesar. Aristeides und Marcus Cato. Perikles und Fabius Maximus. Teil 2: Gaius Marius und Alkibiades. Demosthenes und Cicero*. Translated by Konrat Ziegler and Walter Wuhrmann. Berlin: De Gruyter, 2001.

Sapontzis, Steve F. *Morals, Reason, and Animals*. Philadelphia: Temple University Press, 1987.

Sayers, J. G. "A report to an academy: On carnophallogocentrism, pigs and meat-writing." *Organization* 23, no. 3 (2016): 370–386. DOI: https://doi.org/10.1177/1350508416629454.

Singer, Peter. *The Expanding Circle*. Princeton, NJ: Princeton University Press, 2011.

Taylor, Sunaura. *Beasts of Burden: Animal and Disability Liberation*. New York: The New Press, 2017.

Wibye, Johan V. "Reviving the Distinction between Positive and Negative Human Rights." Ratio Juris 35 (2022): 363–382. DOI: https://doi.org/10.1111/raju.12363.

Wolfe, Cary. Animal Rites: *American Culture, the Discourse of Species, and Posthumanist Theory*. Chicago: The University of Chicago Press, 2003.

CHAPTER 4

Technological Innovation(,) and Moral Progress

4.1 From a Speciesist World to a Multispecies Society: How Technological Innovation Might Enable Moral Progress

> Honestly, I don't think speciesism will be easily overcome and the idea of animal rights will soon prevail; on the other hand, achieving world peace or abolishing torture are also not that easy to come by. So what? It is still wrong to kill and mistreat, and it is still worthwhile to do something against it. (Sezgin 2014, 234)

It is getting better. And it is getting worse. We are—and this seems contradictory at first, but is absolutely possible in a capitalist, democratic world—at a point in time of an urgently needed change in humanity's treatment of nonhuman animals and simultaneously at a point in time in which the exploitation of so-called livestock is reaching an unprecedented peak and continues to increase exponentially:

> [An] overall trend is evident: a trend of states offering greater and better protection to animals. But while it is certainly possible to argue that animal welfare legislation is generally getting tougher, that does not necessarily correlate with a better situation for animals in reality. Take, for example, the

D. Schneeberger, *Envisioning a Better Future for Nonhuman Animals*, The Palgrave Macmillan Animal Ethics Series, https://doi.org/10.1007/978-3-031-75332-9_4

> numbers of animals raised, killed and eaten in modern societies. The global per capita consumption of meat has more than doubled between 1961 and 2007, and is expected to double again by 2050. (Cochrane 2010, 2)

The ethical-normative conclusions of animal ethics scholars tend to overwhelm most humans since “accepting animal rights can involve painful and inconvenient changes in one’s personal life: people may not be ready to give up their favorite meat dishes or leather shoes” (Kymlicka and Donaldson 2014, 117). Those advocating for nonhuman animal rights find themselves in a “David and Goliath” situation. The vast majority of humans wish to consume animal products and the animal industry spends millions of dollars on lobbying for their products (Grossman 2019, 522): “If you can imagine a type of animal abuse, chances are high some form of “lobbying apparatus” exists to advocate for its continued existence”.

A highly important opportunity for a considerable reduction in the suffering that we humans inflict on farmed animals is thus the development and establishment of cultivated animal products (cf. e.g. Rorheim et al. 2016; Post et al. 2020). On 5 August 2013, the first cultivated hamburger, developed by a Dutch research team under Mark Post, was publicly presented and tasted in London. (cf. Bouillon 2015). The Austrian nutritionist Hanni Rützler (ibid.) was amongst the people who tasted the burger and reported: “The meat tasted okay. If you leave out how it was created, you could say the burger tasted like any other. But I still think lab meat will have a hard time catching on in Europe because of the pronounced hostility to technology.”

Since cultivated meat is a scientific achievement that could significantly reduce animal suffering, in 2014, the animal rights organization PETA offered a $1 million prize to the researcher who could produce the first cultivated chicken (PETA 2019): “A million dollars sounds like a lot, but it’s a small price to pay for something that could potentially save millions of lives every hour,” said Ingrid Newkirk, founder and President of PETA USA (ibid.). Even if the prize has yet to be awarded, PETA is satisfied: “A lot has happened in in vitro meat research since the prize money was offered. […] We at PETA are pleased to have sparked debate and brought together researchers, generated interest and investment from the food industry and internet millionaires” (ibid.). It is worth noting that amongst those millionaires investing in the development of cultivated meat were Microsoft CEO Bill Gates and Google founder Sergey Brin.

A certain scepticism towards novel, artificially produced food is understandable; in the final analysis, however, the choices humanity makes about what it eats from the range of things edible for humans is a question of tradition, culture and habit. We cannot look for the cause of "excessive killing [...] in the biological nature of the human species, but must seek it in the culture of socialised man" (Hempolinski 1999, 53). When humans are easily able to obtain tasty, cheap, and conveniently available animal products from cultivated production, this could mean the end of the animal industry as we know it. However, without the further development and groundbreaking economic success of cultivated animal products, the abuse of nonhuman animals by humans is unlikely to come to a (quick) end.

In the face of all the suffering we inflict on nonhuman animals, animal ethics is political in the sense that "tolerance becomes a crime when applied to evil" (Mann 1996). Once we have understood the moral catastrophe that is the way we currently treat nonhuman animals, we are required to help. However, each of us is just one human in a mass of billions of others, most of whom currently want to eat nonhuman animals and pay for what is done to them. We are facing enormously powerful systems of industrialized animal agriculture which perpetuate this moral catastrophe. Cultivated animal products could open the floodgates to societal moral progress.

For Western humans, fair consideration of nonhuman animal interests comes at the price of (prima facie ceteris paribus) no longer killing nonhuman animals for animal products, and this, at the present moment, is evidently too high a price to pay for most humans. Just as slavery was banned in the US in 1865 after the invention of the steam engine—which made it much easier to dispense with slave labour—perhaps the economic triumph and establishment of cultivated animal products will be the necessary precursor to "animal liberation." Although research in the field of cultivated animal products is still to a certain extent in its infancy, and there remains considerable potential for improvement in the production of such products (e.g. in terms of costs, nutrient medium, structure, nutrients, taste, etc.) (cf. Rorheim et al. 2016), it is to be hoped that this field will develop significantly and rapidly over time. In view of the fact that cultivated animal products—in addition to significantly reducing animal suffering—may be healthier (entailing, for example, no antibiotics and no multidrug-resistant germs), more ecological, and (eventually, with luck) cheaper, their future economic triumph could be feasible.

Mark Post, Professor of Physiology at Maastricht University, has been working with a team on the development of cultivated beef since 2004.

Just over a decade after starting this research, he estimated in an interview given in 2015 that cultivated meat would be available in supermarkets in five to ten years' time (Post, accessed 7 October 2015). At the time of writing, in 2024, cultivated chicken can already be bought in a restaurant and a supermarket in Singapore. Producers are facing challenges with scaling cultivated animal products due to high production costs:

> Industry executives said that advances in manufacturing had helped several companies cut production costs by as much as 90% from when they first started years back. Eat (sic; Just) and Avant Meats, for example, have cracked the code on replacing fetal bovine serum, found in the blood of fetuses extracted from cows during the slaughter process, with nutrients to produce cultivated meats. Eat Just has made "significant progress" in cutting costs for each chicken nugget from $50 previously but has not said by how much. "It's too high, and it's embarrassing … We lose money every time someone enjoys our cultivated chicken," Eat Just CEO Josh Tetrick said. Huber's sells the Eat Just cultivated chicken meals for about SG$19 ($14). (Daily Sabah 2023)

A decisive advantage of the economic establishment of cultivated animal products would be that the human potential to make just, empathic, less speciesist decisions and, if necessary, to go against the flow, would no longer be required (at least, not to the degree it currently is). Even if humans continue to look out for themselves and their interests first, the animal industry would—it is hoped—undergo a significant transformation, thanks to this technological achievement.

It remains to be seen if and when cultivated animal products will actually achieve the breakthrough that Mark Post and his colleagues (and investors) foresee. It will also be interesting to see how—and in what measure—the animal industry lobby will fight against it as soon as it realizes both the potential and the smouldering threat of cultivated animal products. Meat producers may simply switch to cultivated animal products or grain or vegetable production, just as many former profiteers of institutional slavery in the US had to find other means of income after abolition. Cultivated animal products could become the game-changer that finally undermines the system more rapidly than animal rights activists and animal ethics scholars could ever have managed. And yet, if we want humans to treat nonhuman animals better, and with more consideration and respect, we must continue to strive not only for technological change, but also for a moral change of heart. In the next section, I discuss four pillars upon which nonhuman animal rights can be built: empathy, justice, a grasp of the inherent value of nonhuman animal life, and knowledge. Using anti-speciesist education, we can strengthen each pillar on an individual level and thereby compliment necessary systemic changes with change efforts on the micro and meso levels. Whilst I am convinced that systemic, institutional change is the only way to free nonhuman animals in the animal industry from their misery, it may still be beneficial to examine other factors that can promote an anti-speciesist change.

4.2 Four Pillars of Nonhuman Animal Rights: Empathy, Justice, Inherent Value of Life, Knowledge

Universal declarations of nonhuman animal rights can only become a reality in our democratic societies if the majority of humans agree to the content of such a declaration. The backbone of universal animal rights declarations, as envisaged here, is the right to life. Inextricably linked to this right to life is the prima facie ceteris paribus injustice of humans living in developed Western societies killing nonhuman animals for the purpose of obtaining animal products. For most humans who hold this view, the animal ethics conviction that this is an injustice rests on several pillars. The first (and perhaps strongest) two pillars are (in no especial order) empathy and sense of justice. The third pillar is formed by a (strong) intuition about the inherent value, i.e. the dignity, (also) of nonhuman animal life (this notion will be discussed extensively in Sect. 8.2). It is recognised (by Schweitzer) that a nonhuman animal is a being who wants to live; that for a living being, its own life is all that it has; that it is a priceless, irrecoverable value. As I argue in Sect. 8.2, the inherent value of a life or a living being can be derived from the striving of the being. The being strives (with all its powers) for its life and flourishing because this is something (perhaps even infinitely and unmeasurably) valuable to the being. The fourth pillar that can support the right of nonhuman animals to their life is formed by the knowledge of the human being(s). Knowledge in the domains of vegan diets, conditions in the animal industry, and the negative ramifications of this industry in terms of world hunger, pollution, rainforest destruction, resource depletion, and climate change, can be beneficial for incrementally transforming our nutritional system.

Knowing that a balanced human diet does not need to include animal products in order to be healthy in all phases of life (Craig and Mangels 2009, 1282) might be crucial for humans in terms of transitioning to a plant-based/vegan diet in the first place. Knowing about the discrepancy between the circumstances and conditions of industrial meat production and the needs of the nonhuman animals (ab)used might help humans to move towards plant-based diets: "A person who wants to eat meat cannot be aware of all the conditions of livestock farming. If they were, they might not be able to eat a single bite" (Benz-Schwarzburg 2014). Similarly, knowledge of the diverse (initially not primarily animal ethical) environmentally harmful effects of the animal industry can help towards reducing human consumption of animal products. The animal industry is a system which not only generates unbearable and unjust suffering, but also contributes to the fact that humans who live in countries where enough grain could actually be grown to feed all inhabitants are starving to death because the grain grown there is instead exported to wealthy Western countries as feed for farmed animals (Food and Agricultural Organization of the United Nations (FAO) et al. 2021). Furthermore, the animal industry pollutes our environment and is one of the main drivers of climate change, since it emits more greenhouse gases than the entire transport sector on this planet combined; it destroys, in manifold ways, extremely valuable and unique habitats such as rainforests at a rapid pace, and wastes resources such as land, freshwater, and energy (Willett et al. 2019; Caro et al. 2015; Gerber et al. 2013; Heffernan et al. 2013; Román-Palacios and Wiens 2020).

Empathy, justice, a grasp of the inherent value of nonhuman animal life, and knowledge: these pillars can be strengthened on the individual levels. We know that speciesism is learned (Wilks et al. 2021; McGuire et al. 2023):

> Our natural self-confidence, sensitivity, and self-respect are repressed as we learn to withhold respect and kindness from other living beings. Through our daily meals, and through zoos, rodeos, circuses, hunting, fishing, and many other socially approved practices and products, we are injected with a cultural program of violence toward defenseless and enslaved animals, and the ripples of that radiate as exclusivism, speciesism, and the many forms of separateness and entitlement that we see around us. (Tuttle 2014, 21)

Humans can, however, also unlearn speciesism (ibid.), just as with other kinds of discrimination (e.g. racism or sexism). We need to make anti-speciesist education accessible just as we make anti-racist and anti-sexist education accessible. Systemic change enabled by technology should therefore be accompanied with educational measures seeking to develop our empathy and sense of anti-speciesist justice. Once we can appreciate the inherent value of nonhuman animal lives and are knowledgeable about healthy vegan diets, as well as the negative ramifications of the animal industry, we might be able to work towards a just multispecies society more effectively.

In the following section, I explore the perplexing question of why some individuals adopt vegan lifestyles due to ethical reasons, whilst the majority of humans continue consuming animal products without much consideration for the nonhuman animals involved. I highlight the disconnect between those deeply troubled by the treatment of farmed nonhuman animals and those who either ignore or accept it. I discuss controversial analogies to historical atrocities such as the Holocaust or slavery, drawing comparisons between systemic violence inflicted on humans and nonhumans. Such comparisons have sparked controversy and resistance, but also challenge societal norms and prompt moral reflection.

4.3 Moral Outrage and the Holocaust Analogy

> Granted that any practice causes more pain to animals than it gives pleasure to man; is that practice moral or immoral? (Mill 1969, 184)

I keep asking myself why some humans follow a vegan lifestyle out of ethical concern for nonhuman animals, while the majority of (Western) humans do not. "[C]onsuming cheap meat and dairy products at every single meal time is simply the norm, [...] most individuals completely ignore the interests of the animals that provide such food" (Cochrane 2010, 137). Castelló (2022, 48) answers my question by suggesting that "Western language plays a crucial role in distorting the sensory and ethico-political capacities of Westerners, which leads many people to be violent against those who deviate from the paradigmatic human person." Trying to protect their positive self-images, the opposing poles ("human carnivores" vs. vegans) sometimes pathologize "the other." There is even a psychiatric diagnosis for humans who show heightened concern for nonhuman animals: "zoophilpsychosis" (Taylor 2017, 151). What is the reason for the difference in the perception and evaluation of the world around us? What causes these highly diverging ethical views and the resulting different diets or lifestyles? How can it be that the current practice of producing, confining and killing farmed animals for the production of meat and other animal products is almost unbearable to me (and others), but is either successfully ignored and suppressed by most people, or assessed and perceived as unproblematic? This is a question that not only concerns me, but other humans; as Sabine Stegmüller-Lang (2014, 67) notes:

> [E]ven after 25 years of intensive reflections I still don't know what motives lie behind obtuse behavior: is it immoderate hubris, tremendous ignorance and lack of compassion or is it simply a lack of knowledge? Perhaps, even after many years of animal protection work, people are still unaware of the suffering caused to animals. For psycho-hygienic reasons, I opt for the third possibility.

He (Castelló 2022, 251) argues against the idea that "if slaughterhouses had glass walls, everyone would be vegetarian" (McCartney 2010). Castelló (2022, 251) explains:

> [L]anguage distorts our sensory capacities, that is, language literally deprives us of sensing other animals in their full ethical weight. For this reason, seeing the suffering of animals through glass walls might not do: if one looks through glass walls but what one sees is the consumable animal, what would that change? The point is that when the conceptuality that structures our subjectivities does not "recognise another," as Wadiwel acutely argues, "violence [can] follow."

From a rational, impartial point of view, the nonhuman animal's interest in their own life ceteris paribus clearly outweighs a Western human's wish to eat meat and animal products; a fair decision or weighing of interests (by a blind, impartial Justice) must (in this context) be in favour of the nonhuman animal. If one imagines that all beings concerned were under a contractualist veil of ignorance regarding their own species membership, it could well be that the beings would agree that humans should not and must not breed, cage and kill other animals for meat consumption, as this is not necessary for them, their nutrition and their health, but many other members of the contractually agreed group would suffer considerably as a result (cf. Dombrowski 1998).

The suffering of nonhuman animals at the hands of humans is an atrocity which has been compared to the worst events in contemporary human history. As early as 1949, Martin Heidegger compared the Nazi Holocaust to the ways in which humans treat farmed animals in the present: "agriculture is now a mechanised food industry, in essence the same as the manufacture of corpses in the gas chambers and death camps" (Cavalieri 2006, 61). Similar comparisons have also been made by some animal rights activists, including PETA's controversial travelling exhibition "The Holocaust on Your Plate," initiated in 2003, which resulted in a years-long legal process between representatives of the Jewish community and PETA (*The Guardian* 2003). Challenging the sacrosanct human-animal divide (cf. Kim 2011), such comparisons are considered by some to be inappropriate and disrespectful. The Holocaust comparison connects "animal suffering to the leading 'archetypes of barbarism'—trying to prompt a radical reevaluation of who animals are and how they should be treated" (Kim 2011, 316). Kim (ibid., 323) offers a nuanced discussion of PETA's "The Holocaust on Your Plate," coming to the conclusion (ibid., 332) that "PETA's use of Holocaust analogies can be defended on moral grounds but critiqued on political grounds." Kim (ibid., 333) concludes that those who reacted with outrage to the analogy might entertain a zero-sum notion about group advancement which is misleading, and instead calls for advocates from various groups to cooperate and join forces in order to "make possible a reconfiguration of the landscape of social justice struggles on a global scale." Kim (ibid.) cites a quotation by Isaac Bashevis Singer, which was used by PETA in this campaign: "Our grandchildren will ask us one day: Where were you during the holocaust of the animals? What did you do against these horrifying crimes? We won't be able to offer the same excuse for the second time, that we didn't know." The Holocaust comparison is perhaps met with resistance partly also due to our speciesism and how we are used to devaluing nonhuman animal life, which might be illustrated in the reaction of Rabbi Avi Shafran, Director of Public Affairs for Agudath Israel of America, in 2005 to PETA's campaign:

> That philosophy, denying humanity's uniqueness, is beyond hurtful. It is evil … Judaism—and civilized society, which has adopted many of Jewish tradition's ideals—maintains a clear and crucial distinction between the animal and the human … Conflating the two worlds … inherently shows disdain for the specialness of the human being … [P]erhaps one day PETA will have the courage to truly apologize, for its core philosophy, disavow it, and re-enter the civilized world. (Kim 2011, 330)

When reflecting on the Holocaust comparison, it is important to understand its historical context of how Jews and Black humans were devalued using animalization:

> For centuries, Jews and blacks [sic!] were essentialized as intrinsically animal-like—base, lowly, brutish, lacking in reason, vicious, dirty, lustful—so as to justify their subordination, exploitation and extermination. The Nazis frequently used animalization to construct Jews as less than human and unworthy of moral consideration. Nazi propaganda linked Jews to apes, featuring hairy Jewish men seducing and assaulting innocent Aryan girls. (Kim 2011, 329)

It comes as no surprise that "multiple psychological mechanisms link negative attitudes toward animals to the dehumanization of human outgroups" (Kymlicka and Donaldson 2014, 120). Using two examples discussed by Aph Ko in *Racism as Zoological Witchcraft* (2019), Castelló (2022, 86) provides drastic illustrations of parallels between racist and speciesist violence. The following story about the slave leader Nat Turner illustrates this:

> First, on the 11th November 1831 Nat Turner led a rebellion of slaves, and an estimate of fifty-five white people were found dead. As a reprisal Nat Turner "was decapitated, skinned, and then materially repurposed into different items" [...]. Turner's skin was boiled down and, among other things, turned into a liquid that white people ingested for medicinal purposes. Second, and as Tommy Curry explains: While we are familiar with the rape of Black women during slavery, the story of Black male rape by white men and women is often overlooked [...]. Black males were raped, mutilated, and eaten during American slavery [...]. The rape of Black males during slavery was often exercise[d] as the ultimate demonstration of power and punishment [...]. In the institution of slavery, white women had complete control and access to Black male flesh. She was a master, so to speak, and able to determine the Black male slave's life or death... In this world, white women could use Black male slaves for their personal enjoyment without risking their reputations.

Castelló (ibid., 87) states that "Turner [...] was literally consumed and devoured by Westerners. What is most important, however, is that such actions were not the result of mere reprisal. [...] it was by being sovereigns over Black "animal-like" bodies that white people desired to consume Black people."

From a non-speciesist, egalitarian point of view, one might recognize some (especially systematic) parallels between the appalling human violence towards human, and that towards nonhuman beings. We lock up nonhuman animals to exploit them; they are only permitted the freedom of movement and the food we give them. We isolate and castrate them, subject them to painful, invasive experiments that forfeit their lives and their chance to thrive, make them work for us; and when they can no longer do so, when they are too sick or broken, when feeding them is no longer worthwhile for us, or just when it suits the "owner," they are killed. Killing them is the final act of injustice. We violently and concretely disregard the interests of the individuals concerned. But as long as a sufficiently strong and potent resistance is lacking, these practices will continue, or even expand and worsen. The farmed animals who are affected by these abuses show that they are totally and unequivocally against such abusive treatment and against their killing. However, their voices are ignored, not understood, hidden, stifled. Processes of the animal industry, especially those that are most disturbing for consumers and thus most damaging to business (especially slaughter), are consequently hidden as much as possible and take place in secret, and not in the agora.

In the next chapter, I will discuss the development and necessity of declarations of nonhuman animal rights, paralleling the Universal Declaration of Human Rights (UDHR), adopted in 1948. When it comes to nonhuman animal rights, separate declarations tailored to each species need to be devised. I emphasize the importance of interdisciplinary collaboration involving biological, ethological, ethical, legal, and political expertise to develop these declarations. By identifying shared needs and interests among different animal species, we can propose a catalogue of rights that reflects the specific capacities and requirements of each species. Crucially, I advocate for the meaningful participation of nonhuman animals in this process, following the principle of "Nothing About Us Without Us," which has been central in disability rights advocacy. Similar to the historical evolution of human rights, nonhuman animal rights are currently in an early phase of intellectual conception and require broader acceptance and implementation. Acknowledging our anthropocentric limitations, I stress the need for epistemic humility and empathy to understand nonhuman animal experiences and advocate for their rights effectively.

References

Benz-Schwarzburg, Judith. "Tierethikerin über tote Giraffe: 'Zoos produzieren überschüssige Tiere.'" Published February 11, 2014. Accessed July 12, 2015. derstandard.at/1389860004472/Tierethikerin-ueber-tote-Giraffe-Zoos-produzieren-ueberschuessige-Tiere.

Bouillon, Barbara. "In-Vitro-Fleisch – Nahrungsmittel der Zukunft." Accessed February 7, 2015. future.arte.tv/de/thema/vitro-fleisch.

Caro, Dario, Steven J. Davis, Simone Bastianoni, and Ken Caldeira. "Global and regional trends in greenhouse gas emissions from livestock." *Climatic Change* 126, no. 1–2 (2015): 203–216. https://doi.org/10.1007/s10584-014-1197-x.

Castelló, Pablo P. "The Language of Zoodemocracy: Contesting Human Sovereignty Over Animals." PhD thesis. Royal Holloway, University of London, 2022.

Cavalieri, Paola. "The Animal Debate: A Reexamination." In *In Defence of Animals – The Second Wave*, edited by Peter Singer, 54–68. Oxford: Blackwell Publishing, 2006.

Cochrane, Alasdair. *An Introduction to Animals and Political Theory.* New York: Palgrave Macmillan, 2010.

Craig, Winston J, and Ann R. Mangels. "Position of the American Dietetic Association: Vegetarian Diets." *Journal of the American Dietetic Association* 109 (2009): 1266–1282. https://doi.org/10.1016/j.jada.2009.05.027.

Daily Sabah. "Even after green light, lab-grown meat yet to take off in Singapore." Published March 8, 2023. Accessed March 23, 2024. https://www.dailysabah.com/life/food/even-after-green-light-lab-grown-meat-yet-to-take-off-in-singapore.

Dombrowski, Daniel A. "Rawls and Animals." *International Journal of Applied Philosophy* 12, no. 1 (1998): 63–77. https://doi.org/10.5840/ijap19981212.

Food and Agriculture Organization of the United Nations (FAO), International Fund for Agricultural Development (IFAD), United Nations Children's Fund (UNICEF), World Food Programme (WFP), and World Health Organisation (WHO). *The State of Food Security and Nutrition in the World 2021: Transforming food systems for food security, improved nutrition and affordable healthy diets for all.* Rome: FAO, 2021. https://doi.org/10.4060/cb4474en.

Gerber, Pierre J., Henning Steinfeld, Benjamin Henderson, Anne Mottet, Carolyn Opio, Jeroen Dijkman, Alessandra Falcucci, and Guiseppe Tempio. *Tackling climate change through livestock: A global assessment of emissions and mitigation opportunities.* Rome: Food and Agriculture Organisation of the United Nations (FAO), 2013.

Grossman, Joanna. "Political Lobbying for Animals." In *The Routledge Handbook of Animal Ethics*, edited by Bob Fischer, 516–529. London: Routledge, 2019.

Guardian, The. "'Holocaust on a plate' angers US Jews." Published March 4, 2003. Accessed February 20, 2023. theguardian.com/media/2003/mar/03/advertising.marketingandpr.

Heffernan, Claire, Mo D. Salman, and Luke York. "Livestock infectious disease and climate change: A review of selected literature." *CAB Reviews* 7, no. 11 (2013): 1–26. https://doi.org/10.1079/PAVSNNR20120011

Hempolinski, Michael. "Haben wir das Recht, Tiere zu töten? Moralische Aspekte der Tötung von Tieren." In *Tiere ohne Rechte?* edited by Jan C. Joerden, and Bodo Busch, 53–61. Berlin: Springer, 1999.

Kim, Claire Jean. 2011. "Moral Extensionism or Racist Exploitation? The Use of Holocaust and Slavery Analogies in the Animal Liberation Movement." *New Political Science* 33 (3): 311–33. doi:https://doi.org/10.1080/07393148.2011.592021.

Ko, Aph. *Racism as Zoological Witchcraft: A Guide to Getting Out*. New York: Lantern Books, 2019.

Kymlicka, Will, and Sue Donaldson. "Animal Rights, Multiculturalism, and the Left." *Journal of Social Philosophy* 45, no. 1 (2014): 116–135. https://doi.org/10.1111/josp.12047.

Mann, Thomas. *The Magic Mountain*. Translated by Helen Tracy Lowe-Porter. New York: Vintage, 1996.

McCartney, Paul. "'If slaughterhouses had glass walls, everyone would be vegetarian'." Published August 5, 2010. Accessed May 15, 2024. https://www.paulmccartney.com/news/if-slaughterhouses-had-glass-walls-everyone-would-be-vegetarian

McGuire, Luke, Sally Palmer, and Nadira S. Faber. "The development of speciesism: Age-related differences in the moral view of animals." *Social Psychological and Personality Science*, 14, no. 2 (2023): 228–237. https://doi.org/10.1177/194855062210861.

Mill, John Stuart. (1969) "Three Essays on Religion." In *The Collected Works of John Stuart Mill, Volume X: Essays on Ethics Religion and Society*. Edited by John M. Robson, 184–187. London: Routledge, 1969.

People for the Ethical Treatement of Animals (PETA). "In-vitro-Fleisch und Clean Meat: Gibt es Fleisch aus dem Labor?" Last modified January 13, 2019. Accessed February 16, 2015. peta.de/laborfleisch#.VOI4wy6umqE.

Post, Mark J., Shulamit Levenberg, David L. Kaplan, Nicholas Genovese, Jianan Fu, Chris Bryant, Nicole Negowetti, et al. "Scientific, sustainability and regulatory challenges of cultured meat." *Nature Food* 1, no. 7 (2020): 403–415. https://doi.org/10.1038/s43016-020-0112-z.

Román-Palacios, Cristian, and Wiens, John J. "Recent responses to climate change reveal the drivers of species extinction and survival." *Proceedings of the National Academy of Sciences of the United States of America*. 117, no. 8 (2020): 4211–4217. https://doi.org/10.1073/pnas.1913007117.

Rorheim, Adrian, Adriano Mannino, Tobias Baumann, and Lucius Caviola. "Cultured Meat: A pragmatic solution to the problems posed by industrial farming: Policy paper." *Sentience Politics* 1 (2016): 1–14. https://ea-foundation.org/files/cultured-meat.pdf.

Sezgin, Hilal. *Artgerecht ist nur die Freiheit*. Munich: C.H. Beck, 2014.

Stegmüller-Lang, Sabine. *Tierschutz macht Sinn*. Neckenmarkt: novum pro, 2014.

Taylor, Sunaura. *Beasts of Burden: Animal and Disability Liberation*. New York: The New Press, 2017.

Tuttle, Will. "Introduction: The Circles of Compassion Vision." *Circles of Compassion: Essays Connecting Issues of Justice*, edited by Will Tuttle, 19–21. Danvers, MA: Vegan Publishers, 2014.

Wilks, Matti., Lucius Caviola, Guy Kahane, and Paul Bloom. "Children prioritize humans over animals less than adults do." *Psychological Science* 32, no. 1 (2021): 27–38. https://doi.org/10.1177/0146167223121939l.

Willett, Walter, Johan Rockström, Brent Loken, Marco Springmann, Tim Lang, Sonja Vermeulen, Tara Garnett, et al. "Food in the Anthropocene: the EAT-Lancet Commission on healthy diets from sustainable food systems." *The Lancet Comissions* 393 (10170) (2019): 447–492. https://doi.org/10.1016/S0140-6736(18)31788-4.

CHAPTER 5

Developing Declarations of Nonhuman Animal Rights

5.1 Why Do We Need Declarations of Nonhuman Animal Rights and How Could We Develop Them?

The original concept behind this book was to propose and discuss a Universal Declaration on the Rights of Nonhuman Animals analogous to *the* Universal Declaration on Human Rights (UDHR) adopted in 1948. However, it soon became clear that *a* Universal Declaration on the Rights of Animals that encompasses *all* animal species would be difficult to formulate and might not be effective. Instead, a separate declaration needs to be developed for each species (Benz-Schwarzburg 2012, 407). Stucki (2023, 96) seems to be in agreement here, when she states that "animal rights must be differentiated in correspondence with animals' specific capacities, interests, and (functional) needs." After all, the UDHR is also limited to *one* species.

Human rights are protective fences around the most important goods of humans. They are necessary as a means to protect every individual independent of their socio-economic status, gender, nationality, and ability. The UDHR is an international, collective declaration that the protection of the most important goods of human individuals has been agreed upon and is to be guaranteed. It provides individuals with ethical guidelines and orientation. Stakeholders such as nongovernmental organizations can

D. Schneeberger, *Envisioning a Better Future for Nonhuman Animals*, The Palgrave Macmillan Animal Ethics Series,
https://doi.org/10.1007/978-3-031-75332-9_5

point out shortcomings in the legal, corporate, or political domain referring to the human rights violations. The international human animal rights convention (i.e. the UDHR) also enables the European Union or the United Nations to sanction states where human rights violations take place. The UDHR, however, does not have the status of international law, since the General Assembly of the United Nations does not have the ability to create international law; rather, it is understood as an ideal. So far, 193 member states of the United Nations have signed the UDHR whilst three (South Sudan, Palestine, Vatican City) have refrained from doing so (United Nations Foundation 2023). Following the adoption of the UDHR in 1948, both the International Covenant on Civil and Political Rights and the International Covenant on Economic, Social and Cultural Rights were adopted by the United Nations in 1966 (United Nations Treaty Collection, n.d.-a, n.d.).

Nonhuman animal rights declarations could provide similar protection to nonhuman animal individuals.[1] Rights are based on interests, needs and abilities that make for a good and happy life. These diverge among the different animal species, even if interspecies commonalities can be identified. In order to be able to present a proposal for a catalogue of rights for a species, biological, ethological, ethical, legal and political expertise must flow together in an interdisciplinary project. Interdisciplinary teams of experts would have to work on this task for decades, and substantial resources would have to be devoted to such an endeavour. Certain needs and interests can be identified that are shared by many (social) animals, for example, in terms of food, exercise, reproduction, social contact, and meaningful occupation. Therefore, general reflections on these basic needs and the associated goods, abilities and interests are possible without focusing exclusively on a certain species.

We humans would have to act as guardians and representatives of nonhuman animals in drafting nonhuman animal rights declarations. The history of the UDHR proves instructive with regard to procedures to develop and negotiate possible future nonhuman animal rights declarations; specialized historians could provide valuable input here. In order to draft possible future nonhuman animal rights declarations, we would have to

[1] Some think that animal welfare laws are similar to animal rights: "If "rights" are understood to be "legal protection against harm, then many animals already do have rights"" (Grossman 2019, 523). While this does have intuitive appeal, in this book, I argue from the point of view that only humans currently enjoy animal rights.

invest considerably more resources in understanding nonhuman animals, their goods, languages, and behaviours. This knowledge—which we can acquire through watching, listening to, and communicating and interacting with nonhuman animals—needs to guide the drafting of future nonhuman animal rights declarations. We still know (too) little about many nonhuman animal species, their abilities, needs and natural behaviours, although surprising findings about various (nonhuman) animal species are constantly being published. Take primatologist Tetsuro Matsuzawa's finding (2009), for example: Matsuzawa's work with chimpanzees led to the revelation that some (trained, captive) chimpanzees can memorize sequences of numbers in fractions of a second. Chimpanzees were thus able to achieve results in a cognitive memory test that far surpassed the memory and perceptual abilities of humans. Of his findings, Matsuzawa (ibid., 311) writes: "[C]himpanzees can outperform humans in a cognitive task. Many people still hold a kind of naive belief that humans are superior to nonhuman animals in all intellectual domains. This is not true."

This book aims to provide a general overview of the common goods and the needs of nonhuman animals and corresponding rights. First of all, then, it should be clarified which goods are important for which species, and in what degree, and why, and further, why and how which humans in which contexts should grant which specific nonhuman animals corresponding rights to the protection of these goods. Possible answers to these questions can and should be approached in two ways.

First, as already mentioned, the gathering of extensive biological and ethological knowledge is necessary. Due to our still limited knowledge of the nonhuman fauna, we need to practice epistemic humility: "[T]here is so much we do not know about other animals, and much of what we think we know is formed by stereotypical views" (Meijer and Bovenkerk 2021, 50). As much as possible must be understood about the species itself, so that an ethical-normative argumentation can be based on and refer to this knowledge. Since it is easier (for us) to think about what the most important goods of humans are and what it means for humans to be denied these, one might approach an understanding of the goods of nonhuman animals in such a way that one first thinks about what the various goods mean for humans: i.e. why they are (so) important for us. Humans are, after all, an animal species that functions according to the same (or often very similar) biological principles as other animals (i.e. the existence of natural innate biological needs, metabolism, striving to maintain homeostasis, maximizing pleasure and minimizing pain and suffering).

The question therefore arises as to what a certain good represents in the human world, before moving on to consider individual goods in nonhuman animal life. Knoepffler (2010, 92) notes that, in this context "we as humans can only approach nonhuman living beings with our human perspective. It is therefore appropriate to start with our experiences as humans (anthropocentric epistemology [...]) in order to overcome ethical anthropocentrism in this way." Although the perspective of our reflections will always remain a subjective-human one and thus we will not escape *epistemic* anthropocentrism in our lifetime, we must try to transcend our position as humans imaginatively, at least to some extent, in order to be able to have a better overview and judgement of the situation, and at least to move away from *ethical* anthropocentrism,[2] in which other species and their goods tend to be held in low esteem.[3]

Whenever possible, we need to take the wishes of nonhuman animals into account. As with the way in which the Convention on the Rights of the Child, adopted in 1989 (United Nations Children's Fund n.d.) and the Declaration on the Rights of Disabled Persons, drafted by the United Nations (2006), were constructed, beings who are not part of the group in question will need to take over the task of drafting any such declaration. Depending on their abilities, beings who are the target groups of these declarations can and should participate meaningfully. As noted above, the motto "Nothing About Us Without Us" is central in disability rights. It relies on the "principle of participation, and it has been used by disabled people's organisations throughout the years as part of the global movement to achieve the full participation and equalization of opportunities for, by and with persons with disabilities" (United Nations 2004).

We know from the evolution of human rights that a new human right goes through a birth process of three stages before it materializes and is implemented (Stucki 2023): first, there is its intellectual inception, then its gradual reception and consolidation in legal and political arenas, and finally its legal recognition and codification. Stucki (ibid.) states that nonhuman animal rights are currently located between the first and second stage, being (ibid., 56) of the opinion that "[a]nimal rights have passed

[2] (Ethical) anthropocentrism can be defined as a theory or outlook which considers "human beings as the most significant entity of the universe" (*Merriam-Webster* online dictionary, n.d.).

[3] "Human exceptionalism and anthropocentrism operate both by centring humans and their allegedly superior capacities, and by ignoring and belittling the subjectivities, experiences, and capacities of other animals" (Blattner et al. 2020, 18).

the stage of intellectual conception and are currently in the emergence phase." One could argue that there is still a considerable amount of interdisciplinary intellectual nuanced work located in the first stage that needs to be done at this point. In the first instance, this intellectual work proposes a better multispecies future with less discrimination, injustice, and suffering on the side of the nonhuman fauna. We are all influenced by the current Zeitgeist; as Stucki (ibid., 58) notes: "At present, states are heavily implicated in systematic (moral) animal rights violations, either by inflicting or facilitating collectively organized violence against animals." We are embodied beings living in a world in which human domination and exploitation of the nonhuman flora and fauna is commonplace, and our imaginative abilities might be restricted by our situatedness, by what we experience and perceive. Second, we would need to become (very) specific. The big, overarching question we first need to ask and answer—if we wish to develop universal nonhuman animal rights declarations—is: "Which groups of nonhuman animals should why have which rights under which circumstances?" This question is enormously comprehensive and complex, and requires a joint, interdisciplinary effort to answer. Think about all the conflicts that naturally occur between all the beings on this planet who rightfully have moral status.[4] How to solve all those conflicts? What about the predator who wants (and currently still needs) to kill prey animals? What about the interest of the prey animal in not becoming prey?

At this point in time, animal rights are still too far removed from mainstream attitudes and efforts in philosophy and politics. This means that currently, we neither have the democratic will nor the resources to make a concerted effort towards a just multispecies society. Humanity is still largely unaware of its speciesism and does not take the suffering and well-being of nonhuman animals seriously enough in order to make a concerted effort to collectively imagine and then also implement the multispecies utopia that some scholars have started to envision (cf. Donaldson and Kymlicka 2013; Castelló 2022a, b).

If we consider the individual goods at stake in the nonhuman animals rights discussed here, we should reflect on why certain individuals (should) have certain rights to certain goods. In doing so, the most important goods must be identified and understood as far as possible. Following

[4] Moral status is a central concept in the animal ethics discourse. A being has moral status if it bears "direct or independent moral importance" due to the fact that it has interests/welfare that can be affected (cf. DeGrazia 2008, 183).

basic moral philosophical groundwork, it will be necessary to consider legally and politically how these most important rights could be guaranteed. Moral philosophy can offer the scaffolding for legal foundations—take the influence of Immanuel Kant's ethics on legislatures worldwide as an example. In this book, I discuss moral rights. Legal rights then follow or go hand in hand with the (desirable) change in social values in the sense of a democratic bottom-up process.

Important initiatives in the past to work towards nonhuman animal rights include Paola Cavalieri and Peter Singer's Great Ape Project (Cavalieri and Singer 1993, 12f) as well as the Declaration on the Rights of Whales and Dolphins, produced in 2010 by the Helsinki Group of scientists led by Cavalieri. Two other important initiatives in this direction are not discussed further in this book, but should at least be mentioned at this point. An important inter-governmental effort to strengthen the protection of nonhuman animals is the Universal Declaration on Animal Welfare, conceived and proposed in 2000 by the World Society for the Protection of Animals (WSPA), known since 2014 as World Animal Protection, in collaboration with other animal welfare organizations (World Society for the Protection of Animals (WSPA) 2007). Another initiative which stands out in this respect is The Declaration of Animal Rights, which is a project by the planet conservation group Our Planet. Theirs Too (Our Planet. Theirs Too. 2011).

Legal scholars and lawyers have invested considerable effort in making strides towards nonhuman animal rights in the legal domain as well. The goal of many legal efforts in this direction has been to achieve groundbreaking precedent rulings, thereby paving the way towards nonhuman animal rights, similar to how precedent rulings paved the way for the abolition of slavery among humans (Wise 2005). Whilst considerable effort has been made already to work towards nonhuman animal rights becoming a reality, a lot of (anti-speciesist) work lies ahead.

Since nonhuman animal rights could be considered as post-human rights (cf. Stucki), it might serve well to study how the first universal animal rights declaration, the human rights declaration, came to be. In the following chapter, I thus explore the historical evolution of human rights, tracing their development from ancient moral norms to modern milestones. Following the devastation of two world wars, especially the horrors of the Holocaust, safeguards against future atrocities were needed. This led to the adoption of the Universal Declaration of Human Rights in 1948, which aimed to protect basic rights and dignity for all human individuals.

5.2 The Historical Emergence of Human Animal Rights

Moral norms, ideas and rules have accompanied humans ever since they were more nonhuman than human. Various nonhuman social animal species naturally live in social structures shaped by behavioural norms. A landmark in the moral shaping of modern Western human society is the emergence and long-lasting prevalence of Christianity: in the Ten Commandments (adopted from the Jewish tradition), for example, there is a prominent anchoring of the (human) right to life through the prohibition of homicide. In the secular sphere, the French Revolution of the years 1789–1799 AD, with its slogan of *Liberté, Egalité, Fraternité*, particularly highlighted the right to freedom and equal treatment. Recognised entitlement (=positive) rights, respected and protected in modern Western democratic societies are, for example, the right to education or the right to certain social benefits such as basic health care. The American Declaration of Independence (1776), whose Christian influence is apparent, also enshrines the shared moral intuition that every human being is entitled to certain basic rights: "We hold these truths to be self-evident, that all men are created equal, that they are endowed by their Creator with certain unalienable Rights, that among these are Life, Liberty and the pursuit of Happiness" (United States National Archives and Records Administration 2023).

The two world wars of the twentieth century shook the countries affected to the core, bringing with them much death, suffering and poverty; once they were over, one thing was clear: something like this must never happen again. Never again would humanity allow so many humans to be devastated by hardship, suffering and death; never again could humans be deprived of their lives in extermination camps. The horrors brought about by the Second World War in particular were so great that there exists a ban on public expressions of Holocaust-denial in all German-speaking countries, which actually conflicts with the otherwise democratically enshrined freedom of expression. No one may claim under penalty of law that these atrocities did not happen.[5]

[5] Cf. the Austrian Prohibition Act 1947: "§ 3h. According to § 3g, it is also punishable whoever in a printed work, on the radio or in any other medium or who otherwise publicly denies, grossly trivialises, condones or seeks to justify the National Socialist genocide or other National Socialist crimes against humanity in such a way that it becomes accessible to many people" (Rechtsinformationssystem des Bundes (RIS) 2024).

Instead, it is necessary to remember and admonish. Because of the shock felt by humanity, and the horror of what humans could do (and had done) to each other, the demand arose for an internationally adopted, protective fence around the most important goods of every human being: "[T]he non-recognition and contempt of human rights have led to acts of barbarism which fill the conscience of mankind with indignation" (UDHR 1948, Preamble). The international community agreed, therefore, in 1948 that the "equal and inalienable rights of all members of the community of mankind," which accrue to them by virtue of "their inherent dignity," must be recognised and protected collectively (ibid.).

It was not only the Second World War that brought about heinous human rights violations. The widespread public acceptance of slavery in the past, in particular also bears witness to grave moral lapses on the part of humanity. Although slavery was (and indeed still is) present throughout the world, the history of slavery in the US is another especially distressing example of how an entire society can act in a morally reprehensible way. It was not until 1865, almost a century after the proclamation of the Declaration of Independence, that slavery was officially banned in the US. During the time of slavery, the generally accepted average morality was that it was completely unproblematic to own humans as slaves. Even Aristotle, an ancient thinker undoubtedly still influential and considered wise in many ways, was obviously mistaken about his views on slaves, their nature and slavery in general: "By nature, he is a slave who can and does belong to another, and who participates in reason only insofar as he accepts it from others but does not possess it of himself" (Aristotle 2003, v. 1254a). It is striking how similar Aristotle's misjudgement is to that of modern humans about nonhuman animals. We, too, deny them their ability to reason; we, too, assume that we have the right to own nonhuman animals and to determine their lives and fates.

While it is certainly valuable to study how human rights came about, we can also take a look at groups of humans (such as children and disabled humans) who tend to be considerably less powerful and more vulnerable when it comes to fighting for their rights, and who thus need protection and advocacy. Whilst children's rights will appear in passing in the reflections presented in this book, disability rights will feature more prominently. It is highly rewarding to study the important work of critical disability scholars, especially anti-speciesist disability scholars such as Sunaura Taylor (2017), for example.

In the following section, I therefore explore parallels between disability studies and animal ethics, highlighting shared challenges with regard to how marginalized groups are treated. Disabled humans and nonhuman animals used by humans both face disenfranchisement and are unable to fully exercise their abilities and rights within society. The social constructivist view of disability emphasizes society's responsibility to include disabled individuals fully, contrasting with outdated deficit-based approaches. Ableism, i.e. discrimination against disabled individuals, parallels speciesism in devaluing beings based on perceived abilities. The "argument of species overlap" challenges anthropocentric views of value and dignity by comparing nonhuman animals and disabled humans, promoting anti-speciesist intuitions. In examining disability rights, I discuss how the struggles to achieve greater protection for vulnerable groups of humans can inform a movement towards nonhuman animal rights.

5.3 What Animal Ethics Can Learn from Disability Studies

> [D]isability is not a 'lack' but rather 'another mode of existence.' (Castelló 2022a, 143)

The unfair treatment, marginalization, invisibility and lack of recognition experienced by disabled humans has some parallels to the situation of nonhuman animals used by humans. The nonhuman animals concerned might also be seen as "disabled." They tend to be dis-abled in exercising their abilities and satisfying their needs sufficiently and thus in leading a good and happy life. They are (paternalistically) disenfranchised and are barely able, if at all, to defend themselves (sufficiently) against this disenfranchisement. The public macropolitical discourse is denied to them. The modes of communication used in this human political discourse are not accessible to them. They have to rely on representatives from the ranks of the powerful to put their voices and interests on the political agenda. Perhaps it is precisely because of these parallels that learning from disability studies is a worthwhile endeavour when reflecting on nonhuman animal rights.

The terms "disability" and "impairment" have different meanings. An impairment is a medical reality, a disability a social one. This understanding of "disability" is social constructivist in nature, which contrasts with an outdated and obsolete individual-deficit approach that tied disability to the (medical) condition of the individual. Modern disability studies point to the responsibility of society not to disable individuals and instead to include individuals with impairments as full members and participants in society. Disabled humans are therefore not *disabled* themselves, but are *dis-abled* by society from participating in (or being included in) society (Taylor 2017). This approach—which avoids excluding and stigmatizing the individual—goes back to the founders of modern disability studies, the US-American sociologist Irving Kenneth Zola and the British social scientist Michael Oliver. Both developed the social constructivist understanding of disability at the same time in the 1980s (Albrecht et al. 2000).

Parallel to unjust actions or attitudes and prejudices, which can be of a racist, sexist or speciesist nature, disabled individuals are confronted with a special form of discrimination called ableism. In this manifestation of discrimination, individuals with an impairment are sometimes treated with less esteem because of a certain perceived lack of abilities (as compared to "normal" humans and their "normal" abilities) (ibid.). Ableism affects nonhuman animals too, as Taylor (2017) points out. In her book *Beasts of Burden: Animal and Disability Liberation*, Taylor provides examples of how disabled nonhuman animals suffer from human devaluation and abuse. In this book, she recounts the story of a fox who lived with the same disability as she does, arthrogryposis, which is "a rare [...] congenital disorder that is characterized by multiple joint contractures and can incorporate muscle weakness and fibrosis" (Kalampokas et al. 2012). Taylor's disability was caused by US military pollution in the town where she was born. The tap water that her mother drank was—unbeknownst to Taylor's mother—heavily polluted (Taylor 2017). The fox in the story was shot by a human because "it had an abnormal gait and appeared sick" (ibid., 23). However, the fox "had normal muscle mass, and his stomach contained a large amount of digested food, which suggested to researchers that "the limb deformity did not preclude successful hunting and foraging"" (ibid., 23).[6]

Taylor (ibid., 32) also cites an employee at John Morrell & Company, a slaughterhouse in Sioux City, Iowa (closed in 2010), who stated: "The preferred method of handling a cripple at Morrell's is to beat him to death with a lead pipe [...]. It's called 'piping.'" Taylor (2017) elaborates that in

[6] Consider the language used here and the change in pronouns accompanying the ways humans relate to the fox.

industrial farming nonhuman animals have to live in environments that disable all those who are confined there, going on to discuss zoopsychosis, a psychosis nonhuman animals can experience due to the confinement they experience in industrial agriculture. Moreover, breeding methods in the animal industry result in disabilities and genetic makeups that cause suffering to nonhuman beings. Taylor (ibid., 39) provides an example from the Humane Society of the US about fur farming, where foxes, minks, chinchillas, and other species are inbred for specific colours, resulting in "deafness, crippling of limbs, deformed sex organs, screw necks, anemia, sterility, and nervous system disorders."

If we think in new directions in animal ethics—that is, if we are concerned with inclusion, participation and less discrimination based on (ethically non-relevant) characteristics of living beings—the findings of disability studies can be inspiring. We need to strive towards the fair treatment of legally and morally marginalized groups who are dependent on and affected by the decisions of beings who do not belong to such vulnerable groups. When thinking about disability studies and animal ethics, Taylor (2017) reminds us that we need to ask how we can "reconcile the brutal reality of human animalization with the concurrent need to challenge the devaluing of animals."

Taylor (ibid., 58) also draws attention to the fact that "[o]ur anthropocentric worldview is supported by ableism" since both human and nonhuman animals suffer from ableism. Taylor (2017) quotes autistic scholar and animal advocate Daniel Salomon (2010) who states that neurotypicalism privileges a form of cognitive processing and a neurotypical (non-autistic) brain structure, while devaluing other forms of cognitive processing (such as those used by autistic humans and nonhuman animals). Taylor (2017, 58) elaborates: "Speciesism uses ableist logic to function. Presumed to be deficient in human markers of intelligence, animals are understood, to put it bluntly, as stupid." Taylor (ibid., 75) underscores that nonhuman animals "have countless capacities that human beings do not have," and that the "myriad tests humans make to measure abilities, do not necessarily translate across species" (ibid., 77) and goes on to discuss the mirror self-recognition test (also known as the "mirror test" or "mark test," which is also discussed further in Sect. 12.1.1 of this book). This test was developed by the psychologist Gordon Gallup Jr. (1970) in order to test (visual) self-recognition in nonhuman animals. In the test, a mark is put on their forehead of the subjects under study. They are then confronted with a mirror. If they then touch the mark on their head, human observers might then deduce from this that they recognize themselves in the mirror. This test exhibits a speciesist bias, as Taylor (2017, 78) explains:

> For example, gorillas were thought unable to pass the mirror test until scientists realized that gorillas have a strong aversion to eye contact and are easily embarrassed. Gorillas would often leave the mirror, go hide, and then try to remove the mark in private. Joshua Plotnik, the head of elephant research at the Golden Triangle Asian Elephant Foundation in Thailand, commented, "The mark test can be difficult to apply across species because it assumes that a particular animal will be interested in something weird on their body. Primates are interested in such things—we're groomers. But elephants are different. They're huge and they're used to putting things on, not taking things off their bodies, like mud and dirt." This test also privileges vision—we really can't tell much about the fact that dogs, for example, fail the mirror test, because we know vision is not their prominent sense.

The mirror test moreover also exhibits weaknesses concerning cultural diversities among humans: "many non-Western children fail to pass the test even at six years old" (ibid., 77). It therefore also exhibits a Western bias. One could speculate that Western culture might suffer from prioritizing too highly how we look (in the mirror).

Taylor (2017, 61) emphasizes that ableism is not restricted to speciesist thinking, but also present in anti-speciesist arenas, providing the example of animal activists who—with the best of intentions—talk about themselves being "a voice for the voiceless," forgetting that nonhuman animals do have voices, though not human ones. Taylor (2017, 62) provides a beautiful quotation from novelist and activist Arundhati Roy's 2004 Sydney Peace Prize Lecture: "There's really no such thing as the 'voiceless'. There are only the deliberately silenced, or the preferably unheard." Taylor (ibid., 63) comments that nonhuman animals "consistently voice preferences and ask for freedom. They speak to us every day [...]. Animals tell us constantly that they want out of their cages, that they want to be reunited with their families, or that they don't want to walk down the kill chute." We can learn a lot from disability scholars about inclusivity when it comes to voice, agency, and participation.

Another parallel between disabled humans and nonhuman animals is a lack of visibility. The (disabled, disenfranchised) lives and deaths of the nonhuman animals produced and consumed by humans are also invisible:

"We have industrialised animal processing industries designed specifically to hide the relationship between live animals and processed meat" (Schlosberg 2014, 87). A concern shared by disability studies and animal ethics is therefore to make discriminated individuals and their interests visible. In the past, disabled humans were often hidden away; to this day, they are still partly separated from "normal" humans and marginalized. Perhaps because of this separation (also, for example, in separate schools), many humans find it difficult to deal with impaired humans in a normal way; they rarely meet disabled humans and do not know how to treat them in an informal and appropriate way: "Many are indeed quite handicapped in dealing with disabled people" (Fuhrmann 2013, 79). Due to this awkwardness of the "non-disabled," contact with disabled humans tends to be avoided, which in turn results in a certain social invisibility, namely not being noticed and not being (sufficiently) included. The resulting social isolation and exclusion can be a considerable psychological burden for disabled humans. Improved inclusion can counteract the perceived invisibility that many disabled humans experience. An argument in which both the moral status of disabled humans and of nonhuman animals is reflected on is the argument of species overlap: the following section will examine this argument more closely.

5.3.1 *The Argument of Species Overlap*

An argument that has been discussed quite broadly in animal ethics is the "argument of species overlap." This argument is also known under its former name "argument from marginal cases." The argument is now referred to as the argument of species overlap as the intention is to move away from referring to nonhuman animals and impaired humans as "marginal cases." Although this argument is not part of the most recent wave in animal ethics and is seen by some as somewhat outdated, it can still help us access anti-speciesist intuitions and understanding.

This argument had already been advanced in the third century AD by Porphyrios (cf. Dombrowski 1984). While it has been discussed extensively in the animal ethics discourse (for a nuanced discussion, see e.g. Horta 2014), for the purposes of this discussion, we will limit our reflection to its central tenets.

If we (want to) couple the moral status of a being, and thus their rights and the recognition of their value (their dignity), to cognitive capacities and characteristics (measured in human anthropocentric standards), then the case arises that some nonhuman animals would have to be placed in the same category as mentally impaired humans; they would thus have to be accorded (similar or) equal treatment, due to the principle of justice that what is equal must justly be treated equally. If we now want to determine the ethically appropriate fair consideration of these beings in a non-speciesist way, two possibilities arise. Either we extend the rights that are also rightly granted to (mentally) impaired humans in our society to those nonhuman animals who are sufficiently similar to them in terms of the (anthropocentric) abilities under consideration, or we place mentally impaired humans on a par with "equal" nonhuman animals, which would theoretically disenfranchise them (in terms of, for example, the right to life). This second option is obviously unacceptable. But even the first option seems to be accompanied by unpleasant consequences, at least for some humans.

In disability studies, the argument of species overlap is understandably not discussed too readily. The argument can be perceived as misleading or potentially threatening. What is particularly threatening here is the idea that the human rights of humans with impairments could be restricted. Animal ethicists discussing the argument are, of course, concerned with better protecting nonhuman animals and not with depriving impaired humans of the human rights protection that they—like all other humans—are clearly entitled to.

What is problematic about the argument of species overlap—and what is criticized by both animal ethicists and disability scholars—is that behind this way of looking at the status quo is a deficiency perspective. Both nonhuman animals and (mentally) impaired humans are conceptualized as deficient (as compared to non-disabled, "normal" humans). Behind the possible intuitions of this argument may lie a perfectionist anthropocentric attitude: "the 'normal' human being is the most valuable being endowed with the greatest possible dignity." Since the argument of species overlap can thus be criticized as ableist on the one hand and anthropocentric on the other (since it measures by human standards), it thus also certainly has weaknesses and limitations, despite its possible helpfulness in the transition to a less speciesist world.

Those who advocate for the rights and better treatment of disabled humans rightly urge that we must distance ourselves from thinking that the life of a "normal" human being could be worth more than the life of a human being who is disabled (by society). It is therefore not surprising that disability scholars and activists have protested against Peter Singer's theses regarding the comparability of the worth and right to protection of human and nonhuman animal life. Passages like the following are perceived critically by disability scholars:

> No objective assessment can support the view that it is always worse to kill members of our own species who are not persons than members of other species who are. On the contrary, there are [...] strong reasons for believing that taking the lives of persons is intrinsically more serious than taking the lives of non-persons. It seems, for example, that killing a chimpanzee is worse than killing a human being who is not and can never be a person because of a congenital mental disability. (Singer 1994, 156)

In the end, moral growth is about expanding the circle of those who are morally (better) considered and not about discriminating against certain people and pushing them out of this protective circle of human animal rights (= human rights).

In the following chapter, I provide a discussion of concepts such as needs, interests, and capabilities. All those concepts are central to my argumentation since understanding these concepts is crucial for arguing why certain beings should have certain rights. In this chapter, I then also provide reflections on interfering with the needs, capabilities, and interests of sentient beings, potentially altering their fundamental nature. I argue that the pursuit of a good life hinges on satisfying needs, pursuing interests, and actualizing capabilities. Failure to meet needs often results in frustration and suffering, especially in contexts such as industrial animal farming where nonhuman animal needs and interests are frequently disregarded. Scientific advancements such as genetic engineering raise contentious possibilities (cf. so-called blind chicken strategies). I argue that interfering with the needs, interests, and capabilities of nonhuman animals can constitute severe harm to nonhuman animals.

References

Albrecht, Gary L., Katherine D. Seelman, and Michael Bury, eds., *Handbook of Disability Studies*. Thousand Oaks, CA: Sage, 2000.

Aristotle. *Politik*. Translated by Olof Gigon. Munich: dtv Verlagsgesellschaft, 2003.

Benz-Schwarzburg, Judith. *Verwandte im Geist – Fremde im Recht – Sozio-kognitive Fähigkeiten bei Tieren und ihre Relevanz für Tierethik und Tierschutz*. Erlangen: Harald Fischer, 2012.

Blattner, Charlotte E., Sue Donaldson, and Ryan Wilcox. "Animal Agency in Community: A Political Multispecies Ethnography of VINE Sanctuary." *Politics and Animals* 6 (2020): 1–22.

Castelló, Pablo P. "The Language of Zoodemocracy: Contesting Human Sovereignty Over Animals." PhD thesis. Royal Holloway, University of London, 2022a.

Castelló, Pablo P. "The erasures of Peter Singer's theory, and the ethical need to consider animals as irreducible others." *Philosophy Today*, 66, no. 1 (2022b): 637–653. DOI: https://doi.org/10.5840/philtoday2022325454.

Cavalieri, Paola, and Peter Singer, eds. *Menschenrechte für die Großen Menschenaffen: "Das Great Ape Project."* Munich: Wilhelm Goldmann Verlag, 1993.

DeGrazia, David. "Moral Status As a Matter of Degree?" *The Southern Journal of Philosophy* 46 (2008): 181–198. DOI: https://doi.org/10.1111/j.2041-6962.2008.tb00075.x.

Dombrowski, Daniel A. "Vegetarianism and the Argument from Marginal Cases in Porphyry." *Journal of the History of Ideas* 45, no. 1 (1984): 141–143. DOI: https://doi.org/10.2307/2709335.

Donaldson, Sue, and Will Kymlicka. *Zoopolis: Eine politische Theorie der Tierrechte.* Translated by Joachim Schulte. Berlin: Suhrkamp, 2013.

Fuhrmann, Dunja. "'Im Grunde genommen werden doch Menschen mit Behinderung im Leben viel mehr behindert, als sie tatsächlich behindert sind!' Ein Interview." In *Sexualität leben ohne Behinderung: Das Menschenrecht auf sexuelle Selbstbestimmung*, edited by Jens Clausen and Frank Herrath, 77–87. Stuttgart: Kohlhammer, 2013.

Gallup, Gordon G. Jr. "Chimpanzees: Self-Recognition." *Science* 167 (1970): 86–87. DOI: https://doi.org/10.1126/science.167.3914.86.

Grossman, Joanna. "Political Lobbying for Animals." In *The Routledge Handbook of Animal Ethics*, edited by Bob Fischer, 516–529. London: Routledge, 2019.

Horta, Oscar. "The Scope of the Argument from Species Overlap." *Journal of Applied Philosophy* 31, no. 2 (2014): 142–54. DOI: https://doi.org/10.1111/japp.12051.

Kalampokas, Emmanouil, Theodoros Kalampokas, Chrisostomos Sofoudis, Efthymios Deligeoroglou, and Dimitrios Botsis. "Diagnosing arthrogryposis multiplex congenita: a review." *ISRN Obstetrics and Gynecology* 2012, no. 3 264918 (September 2012) DOI:https://doi.org/10.5402/2012/264918.

Knoepffler, Nikolaus. *Angewandte Ethik.* Cologne: Böhlau, 2010.

Matsuzawa, Tetsuro. "Q&A." *Current Biology* 19, no. 8 (2009) 310–312. DOI: https://doi.org/10.1016/j.cub.2009.02.001.

Meijer, Eva, and Bernice Bovenkerk, B. "Taking Animal Perspectives into Account in Animal Ethics." In *The International Library of Environmental, Agricultural and Food Ethics, Volume 33: Animals in Our Midst: The Challenges of Co-existing with Animals in the Anthropocene*, edited by Bernice Bovenkerk and Josef

Keulartz, 49–64. Cham, Switzerland: Springer, 2021. DOI: https://doi.org/10.1007/978-3-030-63523-7_3

Merriam-Webster online dictionary. "anthropocentrism." Published n.d.. Accessed May 7, 2024. https://www.merriam-webster.com/dictionary/anthropocentricism.

Rechtsinformationssystem des Bundes (RIS). "Bundesrecht konsolidiert: Gesamte Rechtsvorschrift für Verbotsgesetz 1947." Published n.d.-a. Accessed May 16, 2024. https://www.ris.bka.gv.at/GeltendeFassung.wxe?Abfrage=Bundesnormen&Gesetzesnummer=10000207.

Roy, Arundhati. "The 2004 Sydney Peace Prize Lecture: Peace and the New Corporate Liberation Theology." Seymour Theatre Centre, University of Sydney. November 4, 2004. https://sydneypeacefoundation.org.au/peace-prize-recipients/2004-arundhati-roy/.

Salomon, Daniel. "Animal Rights and Autism Pride: Let's Heal the Rift." *The Scavenger*, October 2010.

Schlosberg, David. "Ecological Justice for the Anthropocene." In *Political Animals and Animal Politics*, edited by Marcel Wissenburg, and David Schlosberg, 75–89. London: Palgrave Macmillan, 2014.

Singer, Peter. *Praktische Ethik*. Translated by Oscar Bischoff, Jean-Claude Wolf, Dietrich Klose, and Susanne Lenz. Stuttgart: Reclam, 1994.

Stucki, Saskia. *One Rights: Human and Animal Rights in the Anthropocene*. Cham, Switzerland: Springer, 2023.

Taylor, Sunaura. *Beasts of Burden: Animal and Disability Liberation*. New York: The New Press, 2017.

United Nations. "Nothing about Us, Without Us." Published 2004. Accessed March 26, 2024. https://www.un.org/esa/socdev/enable/iddp2004.htm#:~:text=The%20motto%20%E2%80%9CNothing%20About%20Us,and%20with%20persons%20with%20disabilities

United Nations. "General Assembly Adopts Groundbreaking Convention, Optional Protocol on Rights of Persons with Disabilities." Press release. New York: United Nations, December 13, 2006.

United Nations Foundation. "The Universal Declaration of Human Rights turning 75: What you need to know." Published December 6, 2023. Accessed June 05, 2024. https://unfoundation.org/blog/post/the-universal-declaration-of-human-rights-is-turning-75-heres-what-you-need-to-know/.

United Nations Treaty Collection. "Chapter IV: Human Rights. 4. International Covenant on Civil and Political Rights." Published n.d.-a. Accessed March 17, 2024. https://treaties.un.org/Pages/ViewDetails.aspx?src=TREATY&mtdsg_no=IV-4&chapter=4&clang=_en.

United Nations Treaty Collection. "Chapter IV: Human Rights. 3. International Covenant on Economic, Social and Cultural Rights." Published n.d.-b. Accessed March 17, 2024. https://treaties.un.org/Pages/ViewDetails.aspx?src=IND&mtdsg_no=IV-3&chapter=4.

United States National Archives and Records Administration. "America's Founding Documents." Last modified June 5, 2023. Accessed August 12, 2015. https://www.archives.gov/founding-docs.

Universal Declaration of Human Rights. Approved and proclaimed by the General Assembly of the United Nations in 1948. Accessed November 1, 2015. un.org/depts/german/menschenrechte/aemr.pdf.

Wise, Steven M. *Though the Heavens May Fall: The Landmark Trial That Led to the End of Human Slavery*. Cambridge, MA: Perseus Books, 2005.

World Society for the Protection of Animals (WSPA; now World Animal Protection). "Universal Declaration of Animal Welfare." Published 2007. Accessed February 14, 2023. worldanimalprotection.ca/sites/default/files/media/ca__en_files/case_for_a_udaw_tcm22-8305.pdf.

CHAPTER 6

Central Concepts: Needs, Capabilities, Interests, and Rights

When it comes to the question of which rights beings ought to have, we must examine their needs, abilities and interests. A right not to be prevented from satisfying a need is extremely important if the need is strong and its frustration is consequently damaging/painful/unpleasant/torturous. A right to freedom, for example, is important where the need for freedom is urgent (or pressing). A right to freedom of religion is only meaningful where there is also a need for the free practice of religion. For example, according to our current knowledge, denying an earthworm the right to free exercise of religion is unlikely to cause them any harm. For humans, who are potentially religious animals, such a right is nevertheless significant.

6.1 Needs

If we don't feel well, then this may be because at least one need is not fulfilled. Often there is also an accumulation of unmet needs. For example, many humans who are unemployed are relatively quickly threatened by poverty. Unemployment not only jeopardizes the fulfilment of certain elementary biological needs, but may also affect the fulfilment of other psychological needs such as meaningful activity and/or social recognition.

D. Schneeberger, *Envisioning a Better Future for Nonhuman Animals*, The Palgrave Macmillan Animal Ethics Series,
https://doi.org/10.1007/978-3-031-75332-9_6

It is our needs that make themselves felt through unpleasant feelings or sensations when they require satisfaction or fulfilment. Which needs these are is (partly) inevitably dictated by our nature. We cannot choose not to drink, eat and breathe (assuming we want to continue our existence). But we can more easily choose, for example, to live a life as a hermit. Nevertheless, with such a choice, certain social needs will usually inevitably make themselves felt as loneliness envelops the hermit and the unpleasant phenomenological quality of the frustration of social needs seeks to motivate them to seek social warmth and connection.

When our needs are met, we experience satisfaction, we are happy. A good book can stimulate intellectually and thus satisfy intellectual needs (at least temporarily), just as a suitable sexual partner can satisfy sexual needs. A good job not only conveys esteem and recognition, it is ideally also an important source of a meaningful activity, and the income it brings in helps to achieve a standard of living that includes the fulfilment of certain other needs.

Complex beings have many different capabilities, but at the same time also an expansive list of needs. For example, a single-celled organism (possibly) might not have a long list of intellectual or emotional capacities, so the path to a "good" life is probably not as long and arduous for it as it is for more complex life forms. Metabolism, suitable climate, reproduction through simple cell division and the "good" life of a unicellular organism is apparently complete.

Compare the single-cell organism to—to pick an example—a stallion who is kept isolated in a small box without an outdoor exercise area. He suffers more or less from the non-fulfilment of multiple elementary needs. He has a need for species-appropriate foraging, meaningful occupation and free exercise, perception of environmental stimuli, social relationships, sexual activity, etc. Considering the fact that many needs of nonhuman animals kept by humans are insufficiently fulfilled, it is astonishing how resilient many of them are, even in the face of poor housing conditions.

When it comes to improving the situation of nonhuman animals, the question must be posed of how an improvement can be achieved. Such an improvement is achieved when nonhuman animals are able to live happier lives, when their interests are given more attention and when the fulfilment of these is at least made possible more than at the present time. Practically, it is the case that manifold (social) animal species share the same or highly similar important basic interests. They have similar basic needs such as metabolism, sleep, free movement, perception, communication, social contact, reproduction, mating and providing for offspring (compare to this, for humans, the basic steps in Maslow's pyramid of human needs (Maslow 1943), with its physiological needs, safety, and love/belonging.)

In order to grasp how important a certain need is in our lives, we only need imagine what life would be like if we experienced a (complete) deprivation of this need. If we cannot fulfil certain basic needs such as metabolic needs and/or sleep over a certain period of time, we are facing our imminent demise. If (basic) needs are not met, an organism enters a state of stress. The living being suffers or at least experiences (highly) unpleasant sensations/feelings. Through the unpleasant nature of pain and suffering (with the help of the biological programme of pleasure maximization and pain avoidance), sentient animals are motivated to escape this harmful situation or to strive for the fulfilment of their needs: "Pleasure and pain are, in the final analysis, tools: tools by which a living thing capable of experiencing them can ensure its survival and the fulfilment of its needs" (Rollin 2006, 99). Sentient entities are evolutionarily programmed to ensure their homeostasis and wellbeing, their continued existence and the passing on of their genes (if possible). Those individuals who did not have this biological, genetically anchored programme were removed by evolution, they were not evolutionarily successful (enough); that is, they could not ensure their own survival and the survival of any offspring sufficiently well to enable the continuation of their genes.

Besides the most important basic needs such as metabolism and freedom—and beings of different species need freedom in varying degrees and kinds, which I will discuss in more detail in Sect. 11.1—the importance of being able to fulfil our needs should not be underestimated. Many Western humans are used to being able to fulfil the most important basic needs: it is, for example, an important need for us to perceive our environment and to be exposed to various environmental stimuli. Stimulus deprivation is used as a means to torture individuals.[1]

[1] Think about the solitary confinement at the Guantanamo Bay detention camp, for example, where prisoners were forced to live with earmuffs and blindfolds for a (un)certain period of time (sensory deprivation) (cf. Dyer 2002). This type of torture brings about severe suffering and auditory and visual hallucinations.

We might become more aware of how important movement is if we imagine ourselves being tied to a chair. Or if we weren't allowed to leave our bed. A shackled, miserable existence is also possible over a longer period of time, if someone does this violence to us. Thus, we humans in torturous imprisonment would also be (sufficiently) resilient in the sense that we would not die, even though we would suffer considerably and suffer psychological and physical damage as a result, just as the individuals currently in factory farms do, since their ability to move is severely restricted by the conditions in which they are kept.

If we imagine situations in which we would have to do without the fulfilment of the needs described above (e.g. movement, sensory needs), we automatically realize that the person causing these conditions would be committing terrible violence against us were they to tie us down and/or lock us up in a gloomy, isolated, musty room. At the same time, we find ourselves in a world where we humans do exactly this to an excessive number of living beings. We chain cows in dark barns, confine pigs,[2,3] use them to our advantage before finally disregarding their fundamentally most urgent interest—the interest in not being killed—by taking their lives. We are doing to sentient beings exactly what we would condemn as the worst atrocities if they were done to us. Why do we judge the situation so differently when we are not the (nonhuman) victims but the (human) perpetrators? Shouldn't contemporary humans, in accordance with their modern nature, their capacity for empathy and their enlightened sense of justice,

[2] "In Austria there are about half as many pigs as humans, but in everyday life you see many humans and no pigs at all. They are locked away in factory farms that look like industrial halls on the outside. [...] The Animal Welfare Act and its regulations allow fattening pigs weighing up to 110 kilograms to be kept in an area of 0.7 square meters, 1 meter × 70 centimeters, without fresh air, without sunlight, without any intellectual stimulation. [...] if you cram even more pigs into an even smaller space, the mortality rate increases so much that the profits go down. The target of 0.7 square meters per pig is exactly the kind of space that promises to maximize profits, as has been shown after many trials. [...] Straw costs money. Therefore, the pigs do not get any. Instead, they have to live on a slatted floor" (Balluch 2014, 133).

[3] Pigs are intelligent and agile beings—[[The] "[w]ild boar is the wild ancestor of the domestic pig. They generally live in forests and woodlands, and they are rapid runners and good swimmers" (Kornum and Knudsen 2010, 3)]—and are supremely adaptable: "It is well known that *S. scrofa* in general are capable of adapting to and thriving in adverse circumstances and this might be seen as an indication of a high level of behavioural plasticity, which will also make this species capable of solving complex cognitive tasks" (Kornum and Knudsen 2010, 3).

come to the conclusion that nonhuman living beings (at least in part) have needs just as urgent as ours, that they suffer when they cannot fulfil them, and that it is prima facie[4] injustice to deprive someone of their life? (Or: "How unjust may we be against animals without negating our own humanity?" (Herzberg 1999, 33).)

Universal animal rights declarations for nonhuman animals would exemplify the fact that we humans have recognised that our fellow animal beings have the same (or at least sufficiently analogous) important, urgent basic needs and goods as we do. They too want to live, move, be happy, entertain and occupy themselves meaningfully, communicate, love each other, be free. And they suffer when they cannot satisfy these needs. How do we humans come to deny them access to these goods just to secure a comparatively vanishingly important luxury or consumer good for ourselves?

6.2 Interests

It has already become clear that nonhuman animals also have interests. In my view, sentient beings do not have to be aware that they have a certain interest in order to have it de facto.[5] Indicators for the existence of an interest are, firstly, acts of communication of whatever kind by the subject that has this interest, which have the (existence of this) interest as their content, and secondly, actions that reveal interests (cf. "voting with one's feet"[6]). A nonhuman animal can cast their vote (also potentially understandable to humans) with their actions without using human written or spoken language. This is not only clear in the case of nonhuman animals, but also in the case of humans with impairments who communicate their interests, wishes and opinions in non-verbal language. However, in the interaction between human and nonhuman animals, a decent knowledge of the species or perhaps even of the individual is a prerequisite for being able to correctly interpret nonhuman animal

[4] The human right to life is an example of a prima facie right, as there are circumstances that can undermine it (e.g. self-defence).

[5] Cf. also Jean-Claude Wolf's (1990, 221; 2000, 44) distinction between subjective and objective interests, according to which a state of affairs can be in the objective interest of a being, even if this being is not subjectively aware of this interest.

[6] Cf. e.g.: "[…] [W]hen worms' lives are under threat by predators such as moles, they literally run for their lives! They escape and move faster than usual" (Castelló 2022, 172).

language. One could argue that interests in the broadest sense are not only recognizable in animals, but also in plants. (I will discuss this aspect in greater detail in Sect. 8.3.1.)

The argumentation presented here is based not only on needs and capabilities, but also on interests. Interests are a biologically well-founded concept and therefore suitable in animal ethics when reflecting on nonhuman animal rights (cf. also Cochrane 2007). This is also the case because interests—assuming a rather broad concept of interests—are relatively easy to recognize once one understands the biology and language of nonhuman animals sufficiently well and can empathize with them to a certain degree. Interests thus form a solid (since biological, empirically perceptible and not metaphysical) basis for a normative argumentation.

It is clear, for example, that a dog who is being held down (by an adult human) and tortured for the torturer's sadistic pleasure, who squirms, and wants to escape their painful mistreatment, possibly communicating their pain through vocalizations, has an urgent interest in freeing themselves from this painful situation and thereby protecting themselves, their welfare, health and life. The dog has an obvious interest in not being tortured. Humans are usually empathically affected when they witness violence towards animals. For empathic beings such as us humans, it is a normal impulse to want to come to the dog's aid: "Mirror neurons as a basic tool of social communication [...] form the basis for the 'empathic impulse' [...] some mammals including humans spontaneously rush to the aid of others without this being associated with an immediate advantage for the aid provider" (Kotrschal 2013, 337). Another normal consequence of observing such an act is that a certain fear (and a certain antipathy) arises in the observer towards a person who carries out such torture, because the assumed abnormal psychological nature of the torturer exemplified here can also be potentially dangerous for witnesses who are initially not directly involved. Obviously, sadistic cruelty to nonhuman animals is considered to be an abnormal human behaviour that can be symptomatic of a psychological disorder.

We tend to value the interests of an individual as more important (to us) the closer the being that is the bearer of the interests is to us. We are (at least vaguely) aware that children in, say, parts of Africa starve to death every day or die of diseases that could be easily cured with medicines that are readily available to Western humans. But this fact is not enough for the vast majority of us to become actively involved in ameliorating their situation. If we have to watch our own child wasting away and/or being tortured, we suffer through our capacity for empathy and emotional connection with the child: "To love someone means [...] that my wellbeing depends on to the wellbeing of the beloved person" (Sedmak 2012, 35). The fact that we are more affected by suffering which we perceive directly and that affects someone close to us than by suffering of which we are aware, but which takes place without our direct perception, is part of why this world includes such terrible atrocities as the current animal industry. This empathic and altruistic fixation on our connection with beings (similar to us) in our personal proximity is (biologically) meaningful: no one could carry the entire suffering of the world on their shoulders, but everyone can (to a certain extent) take care of the beings in their more proximate social structure whose wellbeing and suffering they can perceive and understand.

The question of which nonhuman animals should have which rights can only meaningfully be answered in the light of biological knowledge about the existing interests, needs and abilities of a being. For many nonhuman animal species, such an understanding on the part of humans is not yet available; however, we do not have to be biologists or ethologists to be able to observe interests. Interests should be perceived, recognised and fairly taken into account.

6.3 Capabilities and Telos

The goal or function of animal rights should be to enable a happy and good life for those animals that are recognised as the bearers of these rights. A good animal life is afforded when animals have sufficient opportunities to satisfy their natural needs and to exercise their species-appropriate abilities and behaviours. If animals are given sufficient opportunities to do this, this is naturally usually accompanied by the animals' satisfaction and their subjective wellbeing (their happiness in life): "[S]atisfaction of telos results in happiness for an animal" (Rollin 2011, 51). Rollin (ibid.) states that every species has a telos (Latin for "goal," "purpose"); the concept of this telos goes back to Aristotle. Rollin (ibid.) casually sums up how he understands the normative implications of a telos: "Fish gotta swim; birds gotta fly." Rollin (2006, 99f) illustrates his argument with a comparison between a stone and a spider: in contrast to a stone, "when the spider is alive, it has what Aristotle called a telos, a nature, a function, a set of activities intrinsic to it, evolutionarily determined and genetically imprinted, that constitute its 'living spiderness' [...]

[I]ts life consists precisely of a struggle to perform these functions, to actualize this nature, to fulfil these needs, to maintain this life."

The Aristotelian concept of telos—which has also influenced Nussbaum's (neo-Aristotelian) capability approach—can be illuminating and helpful in the modern animal ethics discourse when reflecting on a good and happy animal life. In addition to its application in animal ethics, it can also be used in ecological arguments. Theoretically, a river, for example, could also be assumed to have a telos (cf. Schlosberg 2007, 149). The goal, the functionality of a river, which constitutes its integrity, can be, for example, to flow into the sea (ibid.). Aristotle's thoughts and ideas can serve as inspiration; however, in modern reflection and argumentation, Darwinian and modern ethological insights form the starting points (Rollin 2006). Rollin (ibid., 121) sees the origin of the telos of an animal species located in the specific genetic code: "[T]his code determines behavioural, psychological, and social aspects of a creature's nature as well as physical ones" (ibid.). According to Rollin (ibid.), the question of which abilities and activities belong to the natural telos of an animal species is not too difficult to answer from an epistemic point of view: "it involves only sympathetic observations of the animal's life and activities. Those of us raised [...] in contexts that bring us in close touch with nature come by such knowledge as easily as we breathe."

The very experiences that we humans deprive farmed animals of would actually be amongst the most valuable, intense, powerful and intrinsically gratifying of their existence. These can be diverse experiences that would actually be elementary components of their biological-genetic teloi. In a better world, where humans have increasingly distanced themselves from their past of disrespect, oppression and exploitation of other species, individuals belonging to species other than human would be allowed to fulfil their telos to a greater extent and would thus be able to experience more happiness and meaning in their lives. Nonhuman interests and the teloi of nonhuman animals are currently obviously oftentimes ignored and disregarded. Without the power to defend themselves sufficiently, nonhuman animals are at the mercy of human species egoism (human speciesism). However, an existing need of an individual that comes with a genetically intended ability that would be (an important) part of that animal's telos should be respected to a greater extent, even if the animal concerned (human or nonhuman) is not able to write letters of complaint in human language or deliberate on them in a court in human words.

6.3.1 Martha Nussbaum's Capability Approach

An approach that has been impactful in ethical discussions and which (as already mentioned) harmonizes well with the concept of telos, as well as potentially contributing to advances in animal ethics, is the capability approach, which was originally developed by Amartya Sen (1989, 2004) and initially discussed with regard to humans and their rights. Martha Nussbaum offered the first animal ethics adaptation of the capabilities approach. This invaluable theoretical advance brings new momentum to the debate and can further advance it (Schlosberg 2014). Nussbaum herself (Nussbaum 2010, 474) is convinced of the potential of the animal ethics capability approach and points out the advantages of her biophilic approach:

> On the basis of intuitions [...], the capabilities approach can, in my opinion, go beyond both the contractualist and the utilitarian approaches. Its very starting point leaves contractualism behind: a profound wonder in the face of the diversity of living beings, as well as the desire for their wellbeing and for a world in which beings of the most diverse kinds can thrive. [...] The goal of this approach is for all living beings to flourish as what they are.

Nussbaum's capability approach is also suitable for an argumentation for (positive) nonhuman animal rights: "That animals have a claim to a broad spectrum of capabilities that are essential for a prosperous life in accordance with their dignity is one of the central convictions of the capabilities approach. Animals have justice-based entitlements" (Nussbaum 2010, 528). Although the formulation used here by Nussbaum (ibid.)—"which are essential for a flourishing life in accordance with their dignity"—could be taken to mean that there could be different degrees or types of dignity varying by species, she acknowledges a certain egalitarianism when she admits: "As it seems, there is no respectable way to deny the thesis that all living beings have the same dignity across species" (Nussbaum 2006, 383).

In *The Limits of Justice: Disability, Nationality and Species Belonging* (Nussbaum 2010, 528f), Martha Nussbaum lists the following ten points (which she later elaborates) that nonhuman animals should be entitled to and that need to be considered and taken into account when it comes to the entitlements of (human and nonhuman) animals: (1) life, (2) physical health, (3) bodily integrity, (4) senses, (5) feelings, (6) practical reason, (7) belonging, (8) other species, (9) play, and (10) control over one's environment.

Nussbaum (2010) sees the advantage of the capability approach in the fact that it regards nonhuman animals as active subjects striving for something. This automatically leads to the idea that they have a right to strive for something; they should not be prevented from doing so and (if necessary) should even be supported in doing so. Here, Nussbaum (ibid., 477) refers to entitlement rights, i.e. positive rights that demand much more from humans than negative rights: "Unlike contractualism, the capability approach leads to direct duties of justice towards animals." These direct duties seem noble: a picture is painted of a beautiful world in which nonhuman animals are seen as subjects who not only receive care from humans, but are even entitled to it. Such claims would probably quickly morally overwhelm current (human-dominated) society, in which even negative nonhuman animal rights are difficult to establish. Nussbaum herself sees difficulties regarding the practical implementation of her approach. She (ibid., 549) is aware that the capability approach is demanding for humans:

> It demands much more from them than the classical theories of the social contract, more also than Rawls expects from the citizens of the well-ordered society, although his expectations are also already quite high. If the three problems I have discussed [concerning rights of humans with disabilities, humans of other nationalities, and nonhuman animals] are to be resolved, people must have a great deal of compassion and goodwill and maintain these feelings over time.

Compared to utilitarian animal ethics approaches, which are often used in welfarist animal protection arguments, and neo-Kantian animal rights approaches (e.g. those advanced by Regan), the capability approach emerges from a Thomistic virtue ethics tradition, or is at least strongly influenced by it (Wissenburg and Schlosburg 2014). Despite this virtue ethics origin, some of Nussbaum's formulations seem Kantian: "Our approach treats animals as subjects and actors, not only as objects of compassion. Unlike utilitarianism, it respects each individual being and refuses to aggregate the good of different living beings and life forms. No being is treated as a means to the ends of others or of society as a whole" (Nussbaum 2010, 477).

The capability approach may be helpful in developing future universal rights declarations for specific nonhuman animals as it lists capabilities that animals should be entitled to actualize. The aim of the capability approach is to ensure that all animals can achieve a minimum level of flourishing and can actualize their innate capabilities in their lives. On the goal of the capability approach, Nussbaum (2010, 478) elaborates: "Its aim is to ensure some central claims, [...] but it is directed at capabilities, not activities, and summarizes these on a short list. [...] It would be [...] best to produce a limited catalogue of political principles aimed at empowerment and protection."

Nussbaum is convinced that the capabilities approach can avoid certain problems of other animal ethics approaches. Like many others, she criticizes utilitarianism for overriding individuals (and their individual rights) where necessary (Nussbaum 2010). Peter Singer, a former preference utilitarian and now a hedonistic utilitarian—(cf. e.g. his remarks in this regard at a lecture at the University of Salzburg in 2016)—criticizes in turn (Singer 2002) the capabilities approach to the extent that, in his opinion, neither Sen nor Nussbaum sufficiently emphasize which capabilities are important and why (ibid.). To solve this problem, Singer (ibid.) offers two possibilities: in his first proposed solution, the capability approach could be constructed as a subcategory of preference utilitarianism. Since Singer originally assumed that a good life of a being is given when their informed preferences[7] are fulfilled sufficiently (ibid.). According to this approach, those capabilities are considered important and good that are necessary to enable the fulfilment of the most important preferences (ibid.). The second possibility for solving this problem is, according to Singer (ibid.) offered by hedonistic utilitarianism which states that those abilities would be most important without which the individual would be less happy.

6.4 Interference in the Needs, Capabilities and Interests and Thus in the Nature of Sentient Beings

The path to a good, happy and contented life leads us along the (long and winding) road of needs, interests and abilities. There are different ways of dealing with needs. The most natural way is probably simply (if possible)

[7] Informed preferences are those preferences that beings have when they have access to and are aware of all the information relevant to their situation.

to satisfy them. The hungry eats, the thirsty drinks and the lonely calls friends. However, a meal, like a social event, does not provide lasting satisfaction. In the evening we will be hungry again, and the next day we may be sitting at home, lonely again. We are dealing with a perpetual cycle of states of not being satisfied and states of satisfaction.

If needs are not met, suffering might ensue. When animal welfare issues are denounced, for example, in the animal industry, it is mostly about the fact that the needs and interests of the animals concerned are not sufficiently accounted for. In contrast, there are also forms of husbandry that enable the actualization of diverse abilities and the fulfilment of many needs, even if these are exceptions to the norm in the current animal industry.

Another—conceivable and now increasingly technologically possible, but ethically highly controversial and potentially problematic—way of dealing with needs or the stress and suffering resulting from their frustration is to eliminate certain needs (and thus also certain interests and abilities). The "disenhancement" of nonhuman animals has meanwhile seen considerable discussion in animal ethics (cf. e.g. Thompson 2008; Palmer 2011; Ferrari 2012). Genetic engineering enables us to intervene more and more in nature. Here, applied ethics has the important task of presenting considerations and arguments in order to ideally prevent (also) future, (possibly) irreversible harm and suffering caused by such interventions.

The logic behind the method of eliminating needs in non-species-appropriate housing conditions is the following: meeting the needs (to prevent suffering) is too cumbersome and costly, ergo unprofitable from a purely economic point of view. If it is a matter of merely combating (or preventing) suffering that arises from the frustration of these needs, one might start directly with the needs (and thus inevitably also with the essential interests and capabilities) and eliminate them. Where there are no (strong) needs, there is no (agonizing) frustration. Keeping horses permanently in small, dark boxes, for example, would affect their quality of life less if they did not have a strong need for exercise, perception and occupation and did not have pronounced social needs. However, if one were to take these needs away, one would probably be removing a considerable part of who they are. A removal of needs could also mean an approximation to an artificial, machine-like or robot-like being. What remains problematic is that in removing needs, one would also be impairing (cap) abilities. If one frees animals from needs (and thus also takes away their abilities), one causes them substantial damage: in fact, one may even destroy their actual nature to a considerable extent.

An artefact of genetic manipulation that veers in this direction is the genetically manipulated blind chicken. In fact, blind chickens have already been created by chance in genetic research. The "blind chicken problem," which arises in the ethical discussion about a possibly morally imperative disenhancement of nonhuman animals, is discussed, for example, by Paul Thompson, Professor Emeritus of Michigan State University, who formerly held the W.K. Kellogg Chair in Agricultural Food and Community Ethics at that institution. Thompson thematized so-called blind chicken strategies in the egg and chicken industry, and discussed whether it would be (possibly) ethically better to produce blind chickens, given the current housing conditions in the US, since they would suffer less in caged conditions. Blind chickens are apparently less likely to pluck out each other's feathers[8] than their sighted (caged) conspecifics. Cannibalism amongst caged chickens could be curbed as a result; as Thompson remarks: "Blind chickens, it turns out, don't mind being crowded together as much as normal chickens do. And so one suggestion is that, 'Well, we ought to shift over to all blind chickens as a solution to our animal welfare problems that are associated with crowding in the poultry industry'" (National Institute of Food and Agriculture (NIFA) n.d.). One problem with this, according to Thompson, would be how to communicate and justify a "blind chicken strategy" to the public (ibid.). If we cause chickens to be born blind, only so that we can keep them even

[8] Feather pecking is thought to be an effect of the frustrated need to wander around and peck at the ground for food:

> By definition, feather pecking is the plucking of feathers from conspecifics. However, these are not only pulled out, but also eaten by the respective chicken. Cannibalism is the pulling and pecking at the skin of other chickens. Plucking can often be observed in hens that have already picked feathers and have thus literally acquired a taste for it. In some cases, cannibalism is also caused by injuries that encourage susceptible animals to peck. In addition, there is also a form of cannibalism that is completely independent of injuries and is limited to very specific parts of the body. For example, as toe cannibalism or cloacal cannibalism. According to the current state of research, the cause of both behavioral disorders is thought to be a kind of displacement activity. [...] It is assumed that it is a substitute satisfaction or reorientation for pecking at food or scratching and pecking at the ground, which the animal was not able to do sufficiently in its younger years. In addition, there are other favorable factors, such as a high stocking density, an unbalanced diet with an insufficient supply of nutrients and generally inappropriate husbandry conditions. (Hühner Haltung n.d.)

more exploitatively for the short time before they are slaughtered, we create another appalling indictment against ourselves, since we would be using nonhuman animals even more selfishly for our own purposes and damaging them severely at the genetic level.

From the "blind chicken solution" one could move (at least hypothetically exaggerating) to a "headless chicken solution":

> In 1993, Robert Burruss wrote an essay in *The Baltimore Sun* in which he predicted that the future of chicken and egg production will include birds "beheaded and hooked up en masse to industrial-scale versions of the heart-lung machines". Since these birds won't move, cages will be obsolete. [...] The naked headless, gutless chicken will crank out eggs till its ovaries burn out. When a sensor senses that no egg has dropped within the last four or six hours, the carcass will be released onto a conveyer, chopped, sliced, steamed and made into soup, burgers and dogfood. (United Poultry Concerns n.d.)

Extrapolating this idea even further would eventually lead to an unconscious cell cluster (consisting of animal protein), which is something that can already be produced. Compared to the ethically problematic case of disenhancing nonhuman beings, cultivated animal products could bring about enormous positive change for nonhuman animals (as discussed in detail in Chap. 4).

In the next chapter, I explore the concept of a just multispecies society. Collaborative efforts are needed to develop new ways of relating across species boundaries which are respectful concerning different ways of being on this planet. We can draw inspiration from vegan animal sanctuaries—such as the LGBTQ-led farmed animal sanctuary VINE ("Veganism Is the Next Evolution"), located in Vermont, USA—which offer prefigurative spaces for imagining an anti-speciesist future. Multispecies ethnography provides an embodied understanding of this spaces and discusses possibilities for nonviolent interspecies coexistence. Interesting questions then arise such as: Where are the borders of our multispecies society? And how should we solve (lethal) conflicts in it?

References

Balluch, Martin. *Der Hund und sein Philosoph: Plädoyer für Autonomie und Tierrechte*. Vienna: Promedia, 2014.

Castelló, Pablo P. "The Language of Zoodemocracy: Contesting Human Sovereignty Over Animals." PhD thesis. Royal Holloway, University of London, 2022.

Cochrane, Alasdair. "Animal Rights and Animal Experiments: An Interest-Based Approach." *Res Publica* 13 (2007): 293–318. DOI: https://doi.org/10.1007/s11158-007-9037-8.
Dyer, Owen. "Prisoners' treatment is "bordering on torture," charity says." *BMJ (Clinical research ed.)* 324, no. 7331 (2002): 187. Doi: https://doi.org/10.1136/bmj.324.7331.187
Ferrari, Arianna. "Animal Disenhancement for Animal Welfare: The Apparent Philosophical Conundrums and the Real Exploitation of Animals. A Response to Thompson and Palmer." *Nanoethics* 6 (2012): 65–76. DOI: https://doi.org/10.1007/s11569-012-0139-1.
Herzberg, Guntolf. "Gerechtigkeit für Tiere." In *Tiere ohne Rechte?* edited by Jan C. Joerden, and Bodo Busch, 29–39. Berlin: Springer, 1999.
Hühner Haltung. "Federpicken." Published n.d.. Accessed September 8, 2015. huehner-haltung.de/gesundheit/federpicken.html.
Kornum, Birgitte R., and Gitte M. Knudsen. "Cognitive testing of pigs (*Sus scrofa*) in translational biobehavioural research." *Neuroscientific Biobehavioural Review*. 35, no. 3 (2010): 437–51. DOI: https://doi.org/10.1016/j.neubiorev.2010.05.004.
Kotrschal, Kurt. "Warum Menschen mit anderen Tieren sozial sein wollen und können." In *Tiere: Der Mensch und seine Natur*, edited by Konrad Paul Liessmann, 333–359. Vienna: Paul Zsolnay, 2013.
Maslow, Abraham. "A Theory of Human Motivation." *Psychological Review* 50, no. 4 (1943): 370–391. DOI: https://doi.org/10.1037/h0054346.
National Institute of Food and Agriculture (NIFA). "Bioethics." Published n.d. Accessed February 4, 2015. nifa.usda.gov/nea/animals/pdfs/bioethics_thompson.pdf.
Nussbaum, Martha C. *Frontiers of Justice: Disability, Nationality, Species Membership*. Cambridge, MA: Harvard University Press, 2006.
Nussbaum, Martha C. *Die Grenzen der Gerechtigkeit: Behinderung, Nationalität und Spezieszuhörigkeit*. Translated by Robin Celikates and Eva Engels. Frankfurt am Main: Suhrkamp, 2010.
Palmer, Clare. "Animal Disenhancement and the Non-Identity Problem: A Response to Thompson." *Nanoethics* 5 (2011): 43–48. DOI: https://doi.org/10.1007/s11569-011-0115-1.
Rollin, Bernard E. *Animal Rights & Human Morality*. Amherst, NY: Prometheus Books, 2006.
Rollin, Bernard E. *Putting the Horse before Descartes: My Life's Work on Behalf of Animals*. Philadelphia: Temple University Press, 2011.
Schlosberg, David. *Defining Environmental Justice: Theories, Movements and Nature*. Oxford: Oxford University Press, 2007.
Schlosberg, David. "Ecological Justice for the Anthropocene." In *Political Animals and Animal Politics*, edited by Marcel Wissenburg, and David Schlosberg, 75–89. London: Palgrave Macmillan, 2014.

Sedmak, Clemens. *Geglücktes Leben*. Freiburg: Herder, 2012.
Sen, Amartya. "Development as Capability Expansion." *Journal of Development Planning* 19, no. 1 (1989): 41–58.
Sen, Amartya. "Capability and Well-Being." In *The Quality of Life*, edited by Martha Nussbaum, and Amartya Sen, 30–53. New York: Routledge, 2004.
Singer, Peter. "Reply to Martha Nussbaum, Justice for Nonhuman Animals, The Tanner Lectures on Human Values." Published November 13, 2002. Accessed June 15, 2015. utilitarian.net/singer/by/20021113.htm.
Thompson, Paul B. "The Opposite of Human Enhancement: Nanotechnology and the Blind Chicken Problem." *Nanoethics* 2 (2008): 305–316. DOI: https://doi.org/10.1007/s11569-008-0052-9
United Poultry Concerns. "The Experimental Use of Chickens and Other Birds in Biomedical and Agricultural Research: Conclusion – Night of the Living Dead for Birds." Published n.d. Accessed February 4, 2015. upc-online.org/experimentation/experimentalConclusion.htm.
Wissenburg, Marcel, and David Schlosburg. *Political Animals and Animal Politics*. London: Palgrave Macmillan, 2014.
Wolf, Jean-Claude. "Singer über Rechte, Recht auf Leben und Euthanasie." *Analyse & Kritik* 12 (1990): 219–225. DOI: https://doi.org/10.1515/auk-1990-0208.
Wolf, Jean-Claude. "Kommunitarismus versus Liberalismus." In *Kommunitarismus versus Liberalismus*, edited by Kurt Seelmann, 37–60. Stuttgart: Franz Steiner Verlag, 2000.

CHAPTER 7

Imagining a Just Multispecies Society—May All Beings Be Happy and Free from Suffering?

Currently, nonhuman animals suffer from human "tyranny" (Donaldson and Kymlicka 2014, 42). Donaldson and Kymlicka (2014, 23f) state that "[j]ustice demands that domination and hierarchy be replaced by relations of citizenship and its accompanying ethos of equality, participation, consent and cooperation." It is clear that on our way to a less speciesist future, we need to make concerted efforts to collaboratively imagine new ways of relating to each other across species boundaries: we need to "develop a new philosophy of interspecies society" (Kymlicka 2022, 213) and work towards a "response-able ethico-politics" (Castelló 2022a, 142) which entails "responsiveness to different ways of living and being" (ibid.) and "seeks to reduce violence as much as possible" (ibid., 148).

When reflecting on our existence, we become aware of how entangled[1] we are: "humans and animals "do not pre-exist their relatings" (Castelló 2022a, 151). Castelló (ibid., 148) elaborates: "As [contemporary ecofeminist scholar Donna J.] Haraway beautifully puts it [...] in When Species Meet, it is important to work 'for the mortal entanglements of human beings and other organisms in ways that one judges, without

[1] Interspecies entanglements are often "constitutively violent" (Castelló 2022a, 147) (cf. also Towne Hirtenfelder and Prouse 2022).

D. Schneeberger, *Envisioning a Better Future for Nonhuman Animals*, The Palgrave Macmillan Animal Ethics Series,
https://doi.org/10.1007/978-3-031-75332-9_7

guarantees, to be good, that is, to deserve a future.'"[2] In a just multispecies society, we engage in "situated zoodemocratic spaces where different animals can co-author politics and policies" (ibid., 151).

Human interventions will be necessary in order to achieve such a utopia and limit unnecessary violence among the species.[3] Vegan animal sanctuaries can provide prefigurative spaces (cf. Maeckelbergh 2011; Monticelli 2018; Reinecke 2018) of the future we wish to work towards. Studying those spaces using posthumanist multispecies ethnography (Pacini-Ketchabaw et al. 2016; Lloro-Bidart 2018; Blattner et al. 2020) can help us imagine possible anti-speciesist futures. In multispeciesist ethnography, we need to "be mindful of the dangers of 'rushing in to interpret and represent' animals, rather than 'holding back, sensing and following'" (Pacini-Ketchabaw et al. 2016, 157). We need to attune "to our own and other bodies [...] not just studying the world through the safety of detached mental processes" (Blattner et al. 2020, 19). The multispecies community at VINE sanctuary can, for example, be highly inspirational for our imaginative endeavours: "In a community like VINE, [...] members of multiple species form a fluid community that regularly integrates newcomers into a relatively stable, secure, free, and complex environment. They share food, spaces, friends, routines, and activities, and must negotiate these interactions in ways that transcend fixed behaviors or species-specific norms" (Blattner et al. 2020, 13). The "norm of carefulness" (ibid., 15) existing at VINE might also be central in future multispecies societies: "VINE is indeed a society whose members learn and negotiate ways of interacting across their differences, enabling a reasonable level of comfort, ease, safety, and companionship for everyone" (ibid., 16). Blattner and colleagues (ibid., 17) go on to state that "living peaceably

[2] It might be important to note that while Castelló has been heavily influenced by Haraway, he (Castelló 2022a, 127) regards her (work) critically as well: "I want to make explicit upfront that as helpful as Haraway's work might be for building a vegan ethico-politics, she has opposed the view that being a feminist entails being a contextual vegan (Franklin and Haraway 2017, 56), and has supported actions such as producing mice with cancer for experimental purposes (1997) or killing animals for 'annual departmental feasts' (Haraway 2008, 297)."

[3] Cf. the following observations Charlotte Blattner and her colleagues made at VINE sanctuary: "[T]the chickens don't hesitate to rip apart a frog who makes the 'mistake' of entering their space" (Blattner et al. 2020, 13).

with others requires that we exercise various kinds of self-restraint, awareness and care of others, predictability and intelligibility in our movements and actions, and willingness to compromise."

Castelló (2022a, 43) suggests that we live in a "shared sensory-conceptual space" in which "human and non-human animals feel water as water, see trees as trees, taste carrots as carrots, smell smoke as smoke, or touch mud as mud." He (Castelló, 1) continues: "[W]e speak the same animal language and make decisions, and in many cases political decisions." Castelló (ibid., 10) talks about a "shared animal language". Human and nonhuman animals become "coanimals" who "coanimalise" and form "coanimalities" (ibid., 150). Castelló's neologisms were influenced by "Haraway's constant use of the prefix 'co-' which means 'together'" (ibid.). He (ibid., 149) embraces "like Haraway, the use of the prefix co- because our humanimal existence can only be understood as being a constitutively entangled one," stating that "for this reason, I want to propose the term 'coanimal'. My aim is two-fold: to fracture our anthropocentric grammar and lexicon, and simultaneously affirm an animal-vegan grammar and language." Castelló (ibid., 158) also coins the term "zooterrapolis" and explains: "Zooterrapolis is not only a term. Like her sisters [Donaldson and Kymlicka's] Zoopolis and [Haraway's] Terrapolis, Zooterrapolis is as much about a paradigmatic shift of our ethico-political established order as of our human subjectivities. This term's etymology comprises the Greek zoo (animal), the Latin terra (earth) and the Greek polis (city-state)." Castelló (ibid., 202) defines zooterrapolis as "the political space where animals exist and meet as differentially equal earthlings, that is, as "plain members and citizens of [the] world" (Calarco 2020, 35) and as "siblings" (Robinson 2013, 189). In the zooterrapolis, different ways of experiencing and being in the world are not only respected, but celebrated (Castelló 2022a, 205). Quoting Sue Donaldson (2020, 721), Castelló (2022a, 202) points out that the zooterrapolis is "a deliberative world of showing, not telling."

Being human comes with certain powers and abilities to influence the lives of nonhuman beings. Those abilities, one could argue, entail responsibilities to secure the greatest possible wellbeing of every being on this planet. We firstly need democratic political systems with democratic majorities of humans who have anti-speciesism as one of the values they hold. At present, we are of course still far from such a starting point, which would allow us to collectively strive for anti-speciesist multispecies societies. As an imaginative exercise, however, let us assume that we have reached such a situation and now wish to imagine and create just multispecies societies. At this point, we would still need to invest considerable time and resources in debate. A central question in this debate would be: What interests of which nonhuman animals should be taken into account in what way? How should conflicts be solved?

Even those humans who now, at the time of writing, are anti-speciesist in the sense that they would generally endorse nonhuman animal rights, have diverging views about what exactly such a society could look like. There are areas where anti-speciesist humans would agree on important measures, such as ending factory farming, or transforming agriculture and the food system towards plant-based systems, in order to tackle the suffering and violence towards farmed animals. Less consensus exists about what our responsibility is concerning wild animal suffering, for example. If we come to the conclusion that we should try to reduce wild animal suffering as well, since suffering is bad no matter who is experiencing it, and no matter who or what has caused it, and no matter if we have any kind of relationship with the ones who are suffering, we should take steps to reduce it, if we are in the position to do so (Johannsen 2021), then we need to debate about and agree on interventions we could take to achieve this goal.

Should we strive towards a global multispecies society and engage in "paradise engineering" (Keulartz 2021). Could (and should) we, instead of being "devils" towards nonhuman animals (Schopenhauer 2010, 23), be instead their "gods," protecting them and taking care of their wellbeing? Castelló (2022a, 206) encourages us to "dream big," since "sometimes what we need is not a minimal agenda": "[W]hat if the driving force to produce meaningful change did not reside in looking at the horrors of factory farming, but rather in envisaging beautiful and joyful futures; what if what we needed was to provide imaginaries full of possibilities, […] to create new relationships and communities; futures that we might long for and might be willing to strive for." Envisioning such a future, Castelló (ibid., 245) imagines what he calls "Zoolondopolis" (cf. also Castelló 2022b), a future version of the city we now know as London. Positioning himself imaginarily in a situation in which he tells how Zoolondopolis came to be, he paints this picture of it: "humans harvesting and farming the land, chicken sleeping in trees, cows grazing, pigeons and seagulls flying around, dogs and pigs playing and taking Hyperloop trains, and humans, pigs and dogs splashing water in the new Russell Square ponds." This sounds dream-like. Castelló (2022a, 246) corrects this utopian impression and states that

> Zoolondopolis was far from a utopia, some humans broke the law and continued to dominate and oppress animals through strict training. Others physically harmed animals. The ancient sexist culture of rape had not fully gone, nor had racism and ableism. The world was still a grim place: oppressions remained, resources were scarce for all, many human and non-human animals had died and suffered unimaginably over the last decades, and the devastating effects of climate change and industrial farming meant that many places on Earth were uninhabitable for many animal species.

When Castelló (ibid., 148) states that we should "reduce violence as much as possible" in the envisioned vegan "response-able ethico-politics" (ibid., 142), he seems to focus primarily on human violence towards nonhuman beings. However, nonhuman animals can also of course be violent towards both human and nonhuman animals. It is interesting to think about how we could prevent this kind of often instinctive violence—(consider, for example, the hunting instincts of cats)—in Zoolondopolis. In Zoolondopolis, the nonhuman right to freedom is taken seriously. By law, all households where humans live with domesticated nonhuman animals have to have doors for those nonhuman beings so that they can leave the house/flat whenever they liked (Castelló 2022a, 244). Since human violence could be limited with education and laws, possible nonhuman violence could maybe also be limited by education. However, education might have its limits: think about the example of the chickens and the frog in VINE sanctuary (cf. endnote number 3). We have now stumbled upon intriguingly nuanced discussions in animal ethics. A more extensive discussion of how we could reduce violence and suffering among nonhuman animals living in the wild will take place in Sect. 15.2 of this book.

At this point in this book, my preliminary reflections are now drawing to a close, before we find ourselves in medias res, reflecting on each of the Articles that provide a backbone for many possible future nonhuman animal rights declarations. The order in which the Articles are presented is inspired by the Universal Declaration of Human Rights. I include a discussion of each Article which is a part of the UDHR, and which also makes sense in universal declarations of nonhuman animal rights. I will first provide an overview of the list of Articles which could be central in possible future nonhuman animal rights declarations. In the first subsection of the next chapter, I then discuss the concept of dignity and its application to all (quasi)sentient individuals in relation to their inherent value of life and rights. I reflect on Regan's concept of "subjects-of-a-life" which extends moral consideration beyond humans to include mammals, birds, and fishes. The idea of inherent value, rooted in Kantian philosophy, faces scepticism from logicians and analytical philosophers, who question the objectivity of such values. Richard David Precht advocates for value subjectivism, suggesting that value is conferred by humans rather than being objectively inherent to a being. In contrast to this, I argue that ultimately, the concept of inherent value aligns with biological principles, where life's striving indicates value. Sentient beings share a fundamental interest in life and wellbeing, regardless of species. This perspective challenges biases and emphasizes the ethical imperative to recognize and respect the dignity of all individuals, human and nonhuman.

In my exploration of dignity and inherent value, I extend these principles to encompass all forms of life, including plants. The continuity of vitality across flora and fauna stands in contrast to inanimate objects. I argue that the defences plants employ against threats reflect a form of inherent value deserving acknowledgement, even though they lack animal-like sentience. Although I do not advocate for rights for plants comparable to animal rights, I stress the importance of a respectful attitude towards plant life and its value. This viewpoint aligns with thinkers like Schopenhauer, who acknowledge a "will to live" in both animals and plants. The discussion about plant sentience and moral status raises thought-provoking questions about our ethical responsibilities towards all living beings. As our understanding of the nature of plants evolves, it challenges conventional perspectives on moral consideration and prompts reflections on our interactions with different forms of life. I then provide an argument for why a human diet in which no "quasi-sentient" being was harmed might already be possible and might be ethically superior to a human diet consisting of plants.

References

Blattner, Charlotte E., Sue Donaldson, and Ryan Wilcox. "Animal Agency in Community: A Political Multispecies Ethnography of VINE Sanctuary." *Politics and Animals* 6 (2020): 1–22.

Calarco, Matthew. *Beyond the Anthropological Difference*. Cambridge: Cambridge University Press, 2020.

Castelló, Pablo P. "The Language of Zoodemocracy: Contesting Human Sovereignty Over Animals." PhD thesis. Royal Holloway, University of London, 2022a.

Castelló, Pablo P. "The erasures of Peter Singer's theory, and the ethical need to consider animals as irreducible others." *Philosophy Today*, 66, no. 1 (2022b): 637–653. DOI: https://doi.org/10.5840/philtoday2022325454.

Donaldson, Sue. "Animal Agora: Animal Citizens and the Democratic Challenge." *Theory and Practice* 46, no. 4 (2020): 709–735. DOI: https://doi.org/10.5840/soctheorpract202061296.

Donaldson, Sue, and Will Kymlicka. "Unruly Beasts: Animal Citizens and the Threat of Tyranny." *Canadian Journal of Political Science / Revue Canadienne de Science Politique* 47, no. 1 (2014): 23–45.

Franklin, Sarah, and Donna J. Haraway. "Staying with the Manifesto: An Interview with Donna Haraway." *Theory, Culture & Society* 34, no. 4 (2017): 49–63. DOI: https://doi.org/10.1177/0263276417693.

Haraway, Donna J.. "FemaleMan©_Meets_OncoMouse™. Mice into Wormholes: A Technoscience Fugue in Two Parts." In *Modest_Witness@Second_Millennium. FemaleMan©_Meets_OncoMouse™*, 49–118. New York: Routledge, 1997.

Haraway, Donna J. *When Species Meet.* Minneapolis MN: University of Minnesota Press, 2008.

Johannsen, Kyle. Wild Animal Ethics: The Moral and Political Problem of Wild Animal Suffering. London: Routledge, 2021.

Keulartz, Jozef. "Should the Lion Eat Straw Like the Ox? Animal Ethics and the Predation Problem." In *The International Library of Environmental, Agricultural and Food Ethics Volume 33: Animals in Our Midst: The Challenges of Co-existing with Animals in the Anthropocene*, edited by Bernice Bovenkerk, and Jozef Keulartz, 99–122. Cham, Switzerland: Springer, 2021. DOI: https://doi.org/10.1007/978-3-030-63523-7_6.

Kymlicka, Will. "Membership Rights for Animals." *Royal Institute of Philosophy Supplement* 91 (2022): 213–244. DOI: https://doi.org/10.1017/S1358246122000078.

Lloro-Bidart, Teresa. "A Feminist Posthumanist Multispecies Ethnography for Educational Studies." *Educational Studies* 54, no. 3 (2018): 253–70. DOI: https://doi.org/10.1080/00131946.2017.1413370.

Maeckelbergh, Marianne. "Doing is Believing: Prefiguration as Strategic Practice in the Alterglobalization Movement." *Social Movement Studies* 10, no. 1 (2011): 1–20. DOI: https://doi.org/10.1080/14742837.2011.545223.

Monticelli, Lara. "Embodying Alternatives to Capitalism in the 21st Century." *TripleC* 16, no. 2 (2018): 501–17. DOI: https://doi.org/10.31269/triplec.v16i2.1032.

Pacini-Ketchabaw, Veronica, Mindy Blaise, and Affrica Taylor. "De-centring the human in multispecies ethnographies." In *Posthuman Research Practices*, edited by Carol Taylor and Christina Hughes, 149–167. Basingstoke: Palgrave Macmillan, 2016.

Reinecke, Juliane. "Social movements and prefigurative organizing: Confronting entrenched inequalities in Occupy London." *Organization Studies* 39, no. 9 (2018): 1299–1321. DOI: https://doi.org/10.1177/0170840618759815.

Robinson, M. "Veganism and Mi'kmaq Legends." Journal of Native Studies 33, no. 1 (2013): 189–196.

Schopenhauer, Arthur. *Senilia – Gedanken im Alter*. Edited by Franco Volpi and Ernst Ziegler. Munich: Beck, 2010.

Towne Hirtenfelder, Claudia, and Carolyn Prouse. "Milking economies: Multispecies entanglements in the infant formula industry." *Environment and Planning E: Nature and Space*, 5, no. 3 (2022): 1296–1318. DOI: https://doi.org/10.1177/25148486211036.

CHAPTER 8

Possible Future Universal Declarations of Nonhuman Animal Rights

8.1 Overview of the Articles

1. **All Individuals Are Born Free and Equal in Dignity and Rights**
2. **Right to Life**
3. **Right to Freedom**
4. **Right to Safety/Physical Integrity; Prohibition of Torture**
5. **Right to Recognition as a Person/(Legal) Subject Before the Law**
6. **Right to a Human Guardian, a Legal Remedy, and an Indirect Right of Legal Action**
7. **Right to Love, Sexuality, Procreation, Pair-Building, Family, and Friendships**
8. **Right to Property**
9. **Right to Macropolitical Consideration and Micropolitical Participation**
10. **Right to Education**
11. **Right to Work**
12. **Right to an Adequate Standard of Living and to Certain Social Benefits (Protection in the Case of Unemployment and Disability, Entitlement to Pension)**

D. Schneeberger, *Envisioning a Better Future for Nonhuman Animals*, The Palgrave Macmillan Animal Ethics Series,
https://doi.org/10.1007/978-3-031-75332-9_8

8.2 Article 1: All Individuals Are Born Free and Equal in Dignity and Rights: On the Dignity of ((Non)human) Animals, the Value of Life

AND

8.3 Article 2: The Right to Life

> As a human community, we are still far from assigning more value than the price per kilo to animals, even if they are persons. (Brensing 2013, 216)

Human dignity is crucial to the Universal Declaration of Human Rights. This is because it the basis for the ethical justification of the human animal rights adopted in the UDHR. Since every human being is endowed with the same inherent dignity, each is entitled to the same inalienable (human) rights: "Recognizing that the inherent dignity and the equal and inalienable rights of all members of the human community are the foundation of freedom, justice and peace in the world, the General Assembly proclaims this Universal Declaration of Human Rights" (UDHR 1948, Preamble). In addition to the "dignity" of a human being, the UDHR also remarks on the "value of the human person": "the peoples of the United Nations having reaffirmed and resolved in the Charter their belief in fundamental human rights, in *the dignity and worth of the human person* and in the equal rights of men and women" (ibid.; my emphasis).

Now, it is not entirely clear what exactly is meant by "dignity" (in the universal declaration on Human Rights), nor how one can tell whether an individual is endowed with dignity or not. In the UDHR, the fact that a being is a human being is the sufficient necessary condition to render them a being with inherent, inalienable dignity. Certainly, it is practical and meaningful to assume the existence of a dignity, a value, if such a concept of dignity is suitable for ensuring the protection of the most important human goods and the dignified treatment of human beings, thereby increasing the chances of a dignified existence for each and every human individual. The concept of human dignity is, after all, ancient and deeply rooted in Western culture and widely accepted. Nevertheless, it smacks of speciesism to assume (as a human being) that humans have dignity (i.e. value) while other, nonhuman beings are denied it. Why should this be the case?[1]

The fact that we recognize the dignity of humans but (possibly) not that of other animals fits well with the insight that we are (still) inculcated with the Aristotelian (or Christian) image of humans as the "crowning glory of creation," as those who are "more equal" than others (to paraphrase George Orwell's *Animal Farm*) or the "greatest achievement" of evolution. Biologically, such exaltation of humanity is not supported; humankind (as a species) is a branch on the tree of evolution that has grown by chance (through undirected evolution) alongside others. Franz M. Wuketits (2015, 14) writes: "For more than two millennia, man's self-perception was essentially one of self-exaltation." Jane Goodall (1993, 16) has a similar view: "[T]hat non-existent barrier […] between 'man' and 'beast' […] was erected in ignorance, as a result of the arrogant assumption, unfortunately shared by vast numbers of people, that humans are superior to nonhumans in every way." Smith (2014, 62) summarizes—with reference to Markell (2003, 2008) and Plumwood (1993, 2002)—that "the dominant human self has relied upon the separation of nature/culture to achieve an illusion of sovereignty in self-other relations, establishing relations of domination and subordination which disrespect the existence of continuities […] both among humans and between humans and nonhumans." As early as 1894, Henry Salt (Chap. 1; my emphasis) demanded that the artificial exaltation of humans must be overcome: "If we are ever going to do justice to the *lower races*, we must get rid of the antiquated notion of a 'great gulf' fixed between them and mankind, and must recognize the common bond […] that unites all living beings in one universal brotherhood." The fact that Salt writes about the "lower races" indicates that he too—like most of us more than a century later—was influenced by the Aristotelian ranking of species into "higher" and "lower/less developed" animal species. Castelló (2022, 26f) explains why this kind of hierarchical thinking is a trap:

[1] It is interesting to note at this point that some states have already included the dignity of nonhuman animals in their constitution (e.g. Switzerland, Luxemburg). Cf. the following elaboration on the Swiss Animal Welfare Act: "The Act defines the dignity of the animal as its inherent worth, which must be respected when handling it. If any strain imposed on the animal cannot be justified by overriding interests, this constitutes a disregard for the animal's dignity. Strain is imposed on the animal when it is caused any pain, suffering or harm. It must not be exposed to anxiety or humiliation" (Federal Food Safety and Veterinary Office 2020).

> I will not talk of the "higher mammals," or animals who are "complex," "sophisticated," have "cognitive capacities," and so on. This kind of hierarchical language connotes that some beings are, by opposition, simple, unsophisticated, inferior, and do not have human-like cognitive capacities, as if those beings lacked certain abilities and were deprived of being like the human. As this discourse is anthropocentric and ableist, it should be avoided.

The *scala naturae* has an effect on our thinking and understanding of the natural world up to the present day:

> The *scala naturae*, or great chain of being [...], as presented by Charles Bonnet in his *Insectologie* (1745), distributed nature's products, living beings included, from the lowest steps of the ladder occupied by fire, air and water, up to the highest steps hosting monkeys, apes and humans, but there are not lower and higher branches in the evolutionary tree of life. [...] [W]e should not expect in the current professional literature in biology, and evolutionary biology in particular, any survival of scala naturae thinking, or of the corresponding language. These expectations, however, fail to be supported by facts. In the current academic literature, the use of progressionist linguistic expressions, perfectly compatible with a scala naturae worldview, are still quite common to date, apparently replacing the concept of a static ladder of ontological levels with an "evolutionary scale" of progress [...] less to more adapted, less to more complex, less to more evolved—in which humans are commonly self-designed as "highest." (Rigato and Minelli 2013, 1)

With the help of Darwinian insights and research based on them, however, it should be our task to abandon this archaic ranking of species. If one tries to step out of the human perspective a little and strips humanity of machines and gadgets, a different view emerges. From the perspective of an eagle or an insect, humans may look less evolved since they are unable to fly. From a fish's point of view, humans are inferior because they cannot breathe under water. From the bat's point of view, humans are deficient because they cannot perceive and navigate by echolocation. From the jaguar's point of view, humans are creepingly slow; they would lose any race by far. Wuketits (2015, 17) summarizes: "If *Homo sapiens* claims a special position for itself, then—if it could—every other species should do so too."

The human assumption (a rather presumptuous one) that we are special and that we and our lives are especially valuable, and that for this reason we deserve special rights and privileges that "the others" do not deserve, seems dubious. Is it not actually evident, precisely because of this limited understanding and the lack of appreciation towards the more-than-human that we may not be as intelligent and wise as we think we are?

For the argumentation presented here, I will now elaborate on my understanding of “dignity.” “Dignity” is normally understood (as it is in this book) as the inherent value of a living being or their life. It is not (in this ethical argument) a question of how valuable a life is instrumentally to anyone other than the living being in question. Drawing on Will Kymlicka’s remarks at the Minding Animals Conference in Utrecht in 2012, Sezgin (2014, 106) notes that: “The right to life is in no way based on the value that this life has for others […] Animals have the right to live because their lives are valuable to them.” Furthermore, “each of us is a unique individual, with unique capacities, and we have an inherent dignity regardless of how we compare to others” (Taylor 2014, 147).

Tom Regan’s thinking has been influential when it comes to the inherent value of beings. All animals that are “subjects-of-a-life” are, according to Regan (2006, 48), bearers of an equal inherent value and are entitled to a right to life. Regan (ibid.) emphasizes that the concept of inherent value is not a gradual one, but a categorical one: i.e. living beings either have inherent value or not. All beings who are bearers of inherent value are thus bearers of the same inherent value. In 2004, Regan (243ff) argued that at least every mammal from the age of one year on is a “subject-of-a-life.” Ten years later, the set of possible “subjects-of-a-life” has expanded: “[W]e don’t have to know everything about subjects-of-a-life in order to know that mammals and birds are subjects-of-a-life. And fish perhaps” (Regan 2014). By stating a minimum set, Regan deftly manoeuvers around the (intuitively and practically) difficult-to-answer question of a lower limit of beings that (should) have the same inherent value and consequently a right to life. Although he specifies a minimum set of those who should have a right to life with his subject-of-a-life criterion, the drawing of the circle of protection around the beings who are subjects-of-a-life must remain “a matter of debate”: “When people ask me: ‘Where do you draw the line?’ I always say: ‘I don’t know. But wherever you draw it, draw it with a pencil. Because who knows what you might learn in the coming years’” (Regan 2014).

The concept of inherent value was originally postulated by Kant, but limited by him to human life (cf. Precht 2000, 247f). Although the concept of inherent value is and has been widely accepted intuitively by humans—at least as far as the inherent value of *human* beings is concerned—it has not been without controversy since its introduction into ethics:

> The idea [of an inherent value of certain beings] originated with Kant, admittedly limited to humans as the only rational being. But even in the discussion surrounding Kant, the idea of such an inherent value is extremely controversial. Logicians, analytical philosophers and critical rationalists find it difficult to accept that the centre of morality lies in an unprovable "value"; [...] they do not care that this value, as so-called "human dignity," is the core of every modern constitution. (Ibid.).

Richard David Precht (2000, 366) also discusses the ontological and epistemological status of such values. He doubts that such values are objective in nature and can be perceived objectively:

> Values are not discovered in the forest like mushrooms, but they are [...] bestowed. [...] The beauty and "value" of a cat is a human ingredient. The special joke is to regard this conferred value as if it were inherent in nature, because that is how it is perceived by many people. (The fact that logicians have problems with such values does not mean that feeling values is nonsense. [...]) The "rights" of animals and nature to their preservation, derived from their "value," are also conferred on them by humans.

The intuition that values are inherent in entities objectively—that is, independently of the attribution of value by an observing subject—is, as Precht already suggests, relatively strong in many humans. This may be because—if one follows my reasoning—it can be assumed that there are objectively recognizable good reasons to assume an inherent value of a life or a living being. The striving to live, thrive and reproduce is a biological principle, it is readily apparent everywhere in flora and fauna. Striving for something appears where there is something valuable (for the striving being) to strive for. Ergo: There is an inherent value objectively present in this living being which is also biologically recognizable from the outside. The will to ("eternal") life (of the genes) is an automatic programme. The will to live and to prosper is a biological principle that is there to protect a value; the most important value: namely, one's own life. Sentient beings strive for intrinsically[2] valuable states, i.e. states of flourishing. Their life, flourishing and

[2] Cf. Regan (1984, 263f), who discusses the common distinction between inherent and intrinsic value. Buschbacher (2015, 10) illustrates the difference between inherent and intrinsic value: "Experiences can be intrinsically valuable for an individual. For example, an experience of pleasure is intrinsically good for the being that experiences it. [...] [A] being

wellbeing are intrinsically valuable to them. Both human and nonhuman animals "are agents who are towards death" (Castelló 2022, 44). We all are vulnerable mortals with an interest in continuing our lives (ibid.).

8.3.1 But What About Plants?

Plants, too, can be perceived as striving to continue their own existence or to ensure their own flourishing and the continuation of their genes. Plants defend themselves when they perceive a threat (cf. e.g. Baldwin and Schultz 1983; De Wit et al. 2012). They do this, for example, by reacting to beings starting to eat them in such a way that they make themselves significantly less palatable, for example, through rapid spontaneous production of bitter substances, or by emitting certain volatile organic compounds (VOCs), which in turn attract the enemies of their enemies, "summoning" help (ibid.). Terrill and Veit (2024, 6) elaborate: "Interestingly, when certain studied plant species—including tomato plants—undergo stress in conditions of cuts, droughts, and herbivores they 'emit remotely detectable and informative airborne sounds' via their tracheas, as well as volatile organic compounds, which inter alia 'may reveal a pathway of signaling between plants and their environment.'" These findings do not enable us to arrive at the conclusion that plants have a subjective experience (they might not, since they do not possess a central nervous system). They, however, certainly underscore our understanding that there is a difference between a stone and a plant. A plant can be dead or alive, thrive or wither away, while a stone could actually not care less—since it cannot care at all. Plants, like animals, and unlike stones, can be dead or alive. This property of vitality distinguishes living things from inanimate things (cf. Schlosberg 2014; Bennett 2010). One can harm plants by damaging or killing them. Damage involves harming or destroying something of value. For the life, wellbeing and flourishing of both a plant and an animal, it does matter how you treat it.

The evolutionary continuity between flora and fauna goes hand in hand with a continuity of vitality and the will to live (Schopenhauer 2005), the striving to live and thrive. In plants, too, thriving or health can be

whose life is intrinsically better is therefore not of greater inherent value according to Regan." The life of a rich CEO is of equal inherent value to the life of a homeless person, even though the CEO's life may be intrinsically more valuable because it may include more pleasurable experiences (ibid.).

distinguished (biologically) from disease, just as a state of being alive can be discerned from a state of not being alive. The fact that plants try to defend themselves can be interpreted as a sign of striving to live and thrive. Thus, it can be assumed that the plant—although it is not sentient in the way a living animal is—strives for something valuable (to it): precisely, for its life, thriving and the reproduction of its genes.

My concept of dignity is therefore not a metaphysical one, but one rooted in the biology of flora and fauna. My argument for the (recognition of) an inherent and intrinsic value of animals is the following:

1. Animals (actively) strive to live, thrive and reproduce.
2. Striving is only meaningful where there is also something valuable to strive for.
3. Living, thriving and reproducing is inherently valuable for sentient beings.

Every living being embodies a value that is demonstrated by its will to live; compare Schopenhauer's "will to live" (2005) or Albert Schweitzer's dictum (2008, 21): "I am life that wants to live in the midst of life that wants to live." My definition of dignity or inherent worth not only includes (also nonhuman) animals, but also plants. However, in contrast to my conviction that in the animal kingdom, dignity does not come in degrees, when it comes to plants, I agree with Terrill and Veit (2024), who propose a gradualist account of quasi-sentience and partial moral status. I also agree here with Duve (2011, 279): "Even if one assumes that a plant has neither a sense of pain nor other states of consciousness nor a will, it does have—like any other living being—three basic interests, namely: (1) to live, (2) to live well, (3) to reproduce." This should not be interpreted to mean that I would now advocate the idea that plants should have a right to life and that we should therefore not harvest and eat them. Nevertheless, a certain respect for the value of plant life, the will to survive and prosper should be expressed, which suggests an appreciative attitude towards plant life. Here I agree with Schopenhauer (2005, 735), who, as is well known, recognised the "will to live" not only in all animals—"We know that the animal wants, even what it wants, namely existence, wellbeing, life and reproduction"—but also conceded a certain degree of wellbeing and woe to plants. Schopenhauer attributes to them a "weak analogue of consciousness, a shadow of comfort and discomfort" (cf. Volkelt 2014, 216).

However, a plant's right to life and possible further plant rights are opposed, ceteris paribus, on the one hand by our (vital) interest in our food (cf. Milburn 2023) and, on the other hand, by the fact that although plants can perceive their own condition and that of the environment in a certain way, they probably do not have any capacity for suffering or a sense of pain equivalent to an animal's sensitivity to pain. If this were the case, they would be pitiful beings who, in the case of an attack, would have only an attempt at defence and (usually rather sluggish and often not very powerful), but never a chance to escape. In flora and fauna it is generally observable that the ability to feel pain tends to go hand in hand with the ability to move (cf. Peter Singer's comments in this regard at a lecture in 2016 in Salzburg).

Research on plant capacities, sentience, and moral status, however, provides thought-provoking reflections. Terrill and Veit (2024) state that our neglect of plant capacities might be a mistake. They argue for giving plants "the benefit of the doubt." As already mentioned above, according to them, a gradualist account of quasi-sentience and partial moral status may be closer to the ethical truth when it comes to plants (ibid.). They talk about minimal morally relevant interests that plants might have and minimal moral obligations that we might have towards them (ibid.; Terrill 2021).[3] Terrill and Veit (2024, 4) cite an example by David DeGrazia which could help us to test our intuitions towards plants: "Imagine angels who are conscious but, lacking feelings, not sentient, and who have the aim of performing certain actions simply because they are right. Even if they do not feel good upon achieving their aims or bad if their aims are thwarted, they have interests in noninterference." Is it easier for us to give partial moral consideration to these angels than to plants (because we intuitively think that these angels have a somewhat similar shape and biology to humans)?

If we take Tom Regan's advice seriously to draw "the line" of inherent value with a pencil (Regan 2014), we should maybe also use this pencil when it comes to plant sentience. If plants have a quasi-sentience that grounds partial moral status, this could, as mentioned above, lead to ethical dilemmas concerning what to eat even for ethical vegans (Terrill and Veit 2024, 8; Milburn 2023). Castelló (2022, 147) contends that "there is no way to eat and not to kill." If we learn more about plant sentience

[3] Cf. Morris (1998, 191) cited in Garner (2013, 9): "[N]ot everything of moral value has moral standing."

and as a consequence start to include them in our moral circle, we might be faced with the ethical request to eat food made in a way that "no plant was harmed" in the process. It is interesting that such food already exists in the form of protein produced from CO2. The Finnish food tech company Solar Foods (cf. Solar Foods 2023) and the Austrian company Arkeon produce such products (cf. Arkeon n.d.). When thinking about such a diet for humans, the *Star Trek* universe might come to mind. In the *Star Trek* universe,[4] food is produced using replicators which generate matter into any desired form, including food and beverages, by combining subatomic particles at the molecular level. A diet in which no quasi-sentient being was harmed might be ethically preferable to vegan diets.[5]

However, since the argumentation in this book refers primarily to nonhuman animals, a reflection on the moral status and the appropriate moral consideration of plants is not the main concern here. A protection of the vital interests, abilities and needs of living beings belonging to certain other animal species, the recognition of the value (i.e. the dignity) of their lives, as well as the granting of basic rights to other species besides humans alone would be a considerable step that still seems (too) great in our present world. "To a first approximation, it would already be a great success if the visible suffering of animals were to cease; only then would we be embarrassed to do research on imperceptible suffering as well" (Herzberg 1999, 34).

[4] Cf. Castelló (2022, 153), who tells us that

> [a]t the end of Zoopolis, […] Donaldson and Kymlicka endorse Star Trek's ethic "for interspecies contact, coexistence, and cooperation" (2011, 257). They summarize the Federation's "first contact" ethic as follows: "encounters with new 'life forms' should be governed by caution, curiosity, and respect […] guided by an overriding injunction to do no harm" (ibid., 257). The USS Enterprise encounters many different life forms, yet they are all respected in their difference and uniqueness. Even though a no-harm ethico-politics is not possible due to the inherent violence that living entails, and especially given our anthropocentric infrastructure, agriculture, energy production systems and law, we should be led by *Star Trek*'s "injunction to do no harm," and truly "do [our] best to limit" violence against animals (Calarco 2004, 194). "Let's break bread at the table together as animals. Let's make kin with respect, without the sword."

[5] The harm done to ***nonhuman animals*** by farming plant-based food might then again also already be reduced considerably by using (vertical) indoor farming.

Discussing the value of nonhuman animal life, Martin Balluch argues that: "No objective differences in the value of life, in the value of freedom and in the value of the integrity of different animals can be rationally justified. [...] Life is all that any animal has and is therefore subjectively valued infinitely in the sense that nothing else can outweigh the value of life" (Balluch 2005, 274) Balluch (ibid., 273) then makes it clear how valuable every animal's own existence is for itself: "[F]or every creature, its own life means everything." If we take the life of an animal, we take "everything" from them. I disagree with those who claim that a human being loses more when they are killed than, for example, a cat. Some people—such as Robert Garner (2013, 15)—assume that the life of a human is worth more than that of a nonhuman animal, but this evaluation takes place from the perspective of a human being. From the subjective perspective, however, the value of one's own life is the same for every animal, regardless of which species one happens to belong to. This probably also applies to insects. For an insect, too, their own survival is the greatest value. From the insect's point of view, the life of any human being might not be that valuable. Similar comparisons can be made concerning intrinsic values of a life. I do not necessarily agree with John Stuart Mill when he says that it is better to be a dissatisfied human than a satisfied pig (cf. Mill 1863, 14).[6] Neither does Julian H. Franklin: "I do not hold that the quality of human life is better than an animal's, nor do I think that a human life is preferable, all things being equal" (2005, 90). Intrinsically, the contented pig probably has more and longer positive experiences. Their cognitive abilities are considered as being comparable to the cognitive potential of dogs (Marino and Colvin 2015). Pigs also possess an astonishing *joie de vivre*:

> In his book *The Whole Hog*, biologist and Johannesburg Zoo director Lyall Watson writes, "I know of no other animals [who] are more consistently curious, more willing to explore new experiences, more ready to meet the world with open mouthed enthusiasm. Pigs, I have discovered, are incurable optimists and get a kick out of just being". Professor Stanley Curtis of Penn State University found that pigs play and excel at joystick-controlled video games. He observed that they are "capable of abstract representation" and

[6] A fly would maybe think: "Those humans… they have no idea what joy it is to be a fly!" (and for many surprising facts about flies, cf. Balcombe 2021).

> "are able to memorize an icon and remember it at a later date." Pigs are much smarter than dogs, according to the research, and even did better at video games than some primates. Says Dr. Sarah Boysen, Curtis' colleague, "[Pigs] are able to focus with an intensity I have never seen in a chimp." (Jaeckle 2011, 36)

Its own life is generally the greatest value, the greatest good, for every sentient being (although in certain cases the survival/life of particularly beloved beings (e.g. one's own child or partner) or of the collective (e.g. an ant colony) may be the greatest good). Thus, if one wants to compare the inherent value of the lives of various animals, one can compare one greatest possible value with other greatest possible values.

From an external perspective, as already discussed in passing earlier, we nevertheless have strong (misguided, speciesist?) non-species-egalitarian value intuitions, for example, we usually find it significantly worse to take the lives of seven humans than seven fruit flies, or seven chickens. Our current legal system also illustrates the common conviction that if we kill humans, we are doing something much worse. We believe this not only because we assume that the friends and family of the human in question would then also suffer (which may also be the case with fruit flies and chickens), but more particularly because most humans have an intuition about the superior value of human life. We carry an intuition within us that leads us to ascribe different values to beings belonging to different species, and we also make speciesist value judgements among nonhuman animal species (Caviola et al. 2019). We believe that it would probably be worse to kill seven apes than seven fruit flies. The same goes for seven horses. Or seven lambs. Or seven cats. Or seven fishes. We need to critically examine such non-egalitarian (speciesist?) intuitions.

To test our intuitions, let us simply imagine Tom Regan's famous "lifeboat case" (cf. Regan 2004, 351): A lifeboat accommodates five occupants (four humans and a dog). To save the others, one occupant must be thrown overboard and thus killed. Regan (1983), with deep misgivings, since he actually sympathizes with an egalitarian position, decides to sacrifice the dog, and justifies this decision with the assumption that humans would have more opportunities for future fulfilling experiences. Although all five individuals would be harmed by dying, the death of the dog is a lesser harm, Regan argues (ibid.), because death is a harm inasmuch as it forecloses opportunities for satisfaction, and he thinks that the humans would lose more of these opportunities than the dog would lose. Regan says: "Death for the dog [...] though a harm, is not comparable to the harm that death would be for any of the humans" (Barad-Andrade 1992, 115). The persuasiveness of this rationale and the suggestion that it seeks to justify a certain speciesism are open to discussion. If we shift perspectives and change roles, we can test our intuitions further. Let's alter the thought experiment slightly, in that there would now be four dogs and one human in the boat, and the dogs would have to decide who would be thrown off the boat. What would their decision be? Do they, speciesistically like us humans, choose the creature that does not belong to their species, i.e. the human? Or are they convinced by Regan's thesis that humans will supposedly have more opportunities to have fulfilling experiences in the future and must therefore be spared at all costs? Or are they wiser (more egalitarian) and come to the conclusion that they cannot decide, since each being's own life is equally valuable to that being itself? Would you agree with Kymlicka's argument which he stated at the Minding Animals conference in Utrecht in 2012 (Sezgin 2014, 106) that it is not the value that is ascribed from the outside but the attribution of value to a life from the inner perspective that is decisive? What if there are four dogs and a guinea pig in the boat? Does the choice fall on the guinea pig because their life (in the dogs' estimation) is not as valuable as a dog's life? One can either wrap oneself in the beautiful egalitarian cloak of wisdom and come to the conclusion that one cannot decide these lifeboat cases, since every living being or their life (at least from the inner perspective—and it is this perspective that actually matters ethically) is equally inherently valuable. Or one removes the rose-tinted glasses and follows one's unsympathetic speciesist (?) intuition and evaluation, which divides animals into higher and lower beings endowed with higher and lower values (*scala naturae*, anyone?), even if such a division is difficult to justify ethically without being biased, egoistic, and/or anthropocentric (cf. Nussbaum 2006, 383, quoted in Donaldson and Kymlicka 2013, 96). I therefore choose the egalitarian point of view when it comes to the inherent value, the dignity, of animals (and animals' lives).

It is our intuitive, human, and pragmatic inability to actually fully comprehend and practice a strict species egalitarianism that Garner (2013, 16) voices as a criticism of abolitionist animal rights approaches: "[T]he abolitionist strand of animal rights should be rejected as an ideal theory because its species egalitarianism [...] seems to demand too much of human beings

[…]. [I]t requires us to engage in a paradigmatic leap across the species divide jettisoning the anthropocentric culture that has dominated human social and political life."

Regardless of the concrete situation in the sinking lifeboat, one could imagine, along the lines of Precht's alien thought experiment (cf. Precht 2011, 144ff), that aliens begin to settle on Earth. They may have special "superhuman" abilities and be able to read minds, travel through time, etc. Assuming a hierarchy of values of the lives of individuals due to the species they belong to as well, one would probably have to agree that killing seven of these aliens is worse than killing seven humans. We humans would likely protest and try to make it clear that *our* lives are just as valuable to *us* as the aliens' lives are to *them*. The value anyone ascribes to our lives from an outside perspective compared to other (perhaps "higher" or "differently evolved") species is irrelevant to the value of our lives to us (which is what should count in the end). But perhaps the aliens are already so far ahead of us morally that they understand this and also regard the lives of "less evolved" beings as equivalently valuable (for the beings themselves). If one compares the values of the lives from the inner perspective, once again it would be impossible to decide who should be thrown out of the lifeboat that accommodates aliens and humans, because if one compares values from the inner perspectives, there is no hierarchy of values.

It can therefore be summarized that at least all sentient animals are inherently equally (namely infinitely (cf. Weil 2012)) valuable, all possess an equal value, an equal dignity. From this inherent value that a life or living being has, the ethical demand can be derived that we humans should respect this value and not damage or destroy it unless there are more important interests on our side that speak against such protection.

The fact that, for every living being, its own life represents a priceless value is evident from how animals fight for their survival, and how much strength they expend to preserve their lives. When it comes to our survival, we fight to the bitter end (unless, for example, we are pathologically depressed and (therefore) suicidal). Thus, the basis for my argumentation (especially concerning the right to life) is the assumption that for all sentient animals, their own lives and flourishing represent an infinitely great value, which goes hand in hand with an urgent interest in securing and promoting their own lives and flourishing. This value coupled with this urgent interest is (also) evident to me in the nonhuman fauna—even in the swimming bee trying to escape a glass of beer—given that I walk through the world with open eyes, an empathic heart and a reflective mind.

Before moving on to the next chapter, I would now like to briefly address key questions around a nonhuman animal right to life. First: Should all nonhuman animals have a right to life? Second: When is killing a nonhuman animal justified? I would like to provide a few reflections on these questions.

It is first and foremost important (as mentioned before) that basic rights are prima facie rights, meaning that there are certain circumstances that can undermine them. The human right to life is also a prima facie right, as there are circumstances that can undermine it (e.g. self-defence). The right to liberty also has exceptions (e.g. there are certain criminal offences that are punishable by deprivation of liberty). In this sense, fundamental rights are not absolute rights. The same would be applicable to human rights. If a human is attacked by a dog, the human is, for example, allowed to use appropriate means to defend themselves. Here, the principle of non-maleficence (Beauchamp and Childress 2013) can be inspirational. We should try to mitigate harm done when defending ourselves: "As little as possible, as much as necessary."

Following the argumentation provided in this book, all sentient nonhuman animals have an equal inherent value, which ought to be weighed equally, and this grounds the call to protect this value. One might now invoke a principle of reasonableness, which asks humans to take reasonable measures to protect the lives of nonhuman animals. In nature there are, of course, innumerable conflicts arising between different beings, though it is—as mentioned before—unfortunately beyond the scope of this book to provide detailed reflections on whose life should be protected in what form under which circumstances. A principle of reasonableness would allow humans to drive a car and thereby inevitably kill flies on the windscreen; it would allow humans to walk across a meadow and thereby unwittingly step on ants. This principle is needed in order to not overwhelm humans in the task of protecting the lives of nonhuman animals. Even though "our existence is unavoidably violent and involves killing" (Castelló 2022, 147), we still have an ethical duty to make considerable efforts to reduce the amount of destruction and suffering we cause (since we are able to do so and we can reasonably be asked to do so without being overwhelmed).

I would now like to switch to an empirical examination of the right to life, reflecting on why certain species receive greater protection of their right to life in (Austria's) animal welfare laws. I will provide many examples from Austria, since this is the country I've lived in for almost my entire life. It is also an interesting country to examine, due to its comparatively strict animal welfare jurisdictions[7] paired with comparatively high meat consumption. Through legal analysis and societal examples, this chapter critically examines the complexities and inconsistencies within Austria's animal protection laws and their implications for a nonhuman animal right to life. I, for example, discuss the fact that in Austria, dogs and cats are protected by law from being consumed by humans but often legally euthanized.

I will in this section also provide a descriptive outlook on why and how some species might be closer to an actual protection of their right to life than others. In general, certain species receive more protection in several parts of the world due to societal familiarity, such as dogs, cats, and horses. Recognition of individuality and subjectivity could drive enhanced protection for nonhuman animals beyond traditional companion species. The chapter highlights evolving attitudes towards nonhuman animal rights, where species like dolphins and apes garner increased protection due to perceived intelligence and personality traits. I discuss the argument that conservation efforts for endangered species indirectly protect individual lives, fostering a possible movement towards intuitively valuing and protecting nonhuman animal life beyond species preservation.

In this next chapter, I also explore the realities of nonhuman animal lives in the animal industry. The Austrian Animal Protection Act prohibits killing nonhuman animals without reasonable cause but includes exceptions that permit killing them on a vast scale for economic reasons. I critically examine practices such as killing male chicks in the egg industry, which is a practice, as I argue, which actually violates current laws against killing animals for purely economic purposes.

[7]The "Animal Protection Index" (API) provides a comparison of the animal welfare jurisdictions of 50 countries and classifies countries into categories ranging from "A" (the best) to "G" (the worst) (Animal Protection Index n.d.). Currently, no country has received an "A" classification (Animal Protection Index n.d.). Austria and Switzerland (category 'B'), as well as Germany (category "C") have amongst the strictest animal welfare jurisdictions worldwide (Animal Protection Index n.d.).

References

Animal Protection Index (API). "About the Animal Protection Index (API)." Published n.d. Accessed August 19, 2021. https://api.worldanimalprotection.org/about.

Arkeon. "Arkeon." Published n.d. Accessed March 24, 2024. https://arkeon.bio/.

Balcombe, Jonathan. Super Fly: *The Unexpected Lives of the World's Most Successful Insects.* London: Penguin Books, 2021.

Baldwin, Ian T., and Jack C. Schultz. "Rapid Changes in Tree Leaf Chemistry Induced by Damage: Evidence for Communication Between Plants." *Science* 221, no. 4607 (1983): 277–279. DOI: https://doi.org/10.1126/science.221.4607.277.

Balluch, Martin. *Die Kontinuität von Bewusstsein: Das naturwissenschaftliche Argument für Tierrechte.* Vienna: Guthmann-Peterson, 2005.

Barad-Andrade, Judith. "The Dog in the Lifeboat Revisited." *Between the Species* 8, no. 2 (1992): 114–117.

Beauchamp, Tom L., and James F. Childress. *Principles of Biomedical Ethics.* New York: Oxford University Press, 2013.

Bennett, Jane. *Vibrant Matter: A Political Ecology of Things.* Durham, NC: Duke University Press, 2010.

Bonnet, Charles. *Traité d'insectologie, ou observations sur les pucerons/Observations sur quelques espèces de vers d'eau douce qui, coupés par morceaux, deviennent autant d'animaux complets.* Paris: Durand, 1745.

Brensing, Karsten. *Persönlichkeitsrechte für Tiere.* Freiburg im Breisgau: Herder, 2013.

Buschbacher, Anton. "Tierrechte." PhD thesis, University of Salzburg, 2015.

Calarco, Matthew. "Deconstruction is not vegetarianism: Humanism, subjectivity, and animal ethics." *Continental Philosophy Review*, 37, no. 2 (2004): 175–201. DOI: https://doi.org/10.1007/s11007-005-3925-4.

Castelló, Pablo P. "The Language of Zoodemocracy: Contesting Human Sovereignty Over Animals." PhD thesis. Royal Holloway, University of London, 2022.

Caviola, Lucius, Jim A.C. Everett, and Nadira S. Faber. "The moral standing of animals: Towards a psychology of speciesism." *Journal of Personality and Social Psychology*, 116, no. 6 (2019): 1011–1029. DOI: https://doi.org/10.1037/pspp0000182.

De Wit, Mieke, Wouter Kegge, Jochem B. Evers, Marleen H. Vergeer-van Eijk, Paulien Gankema, Laurentius A. C. J. Voesenek, and Ronald Pierika. "Plant neighbor detection through touching leaf tips precedes phytochrome signals." *Proceedings of the National Academy of Sciences of the United States of America* 109.36 (2012): 14705–14710. DOI: https://doi.org/10.1073/pnas.1205437109.

Donaldson, Sue, and Will Kymlicka. *Zoopolis: A Political Theory of Animal Rights.* Oxford: Oxford University Press, 2011.
Donaldson, Sue, and Will Kymlicka. *Zoopolis: Eine politische Theorie der Tierrechte.* Translated by Joachim Schulte. Berlin: Suhrkamp, 2013.
Duve, Karen. *Anständig essen – ein Selbstversuch.* Munich: Goldmann, 2011.
Federal Food Safety and Veterinary Office. "Dignity of the animal." Last modified August 11, 2020. Accessed March 26, 2024. https://www.blv.admin.ch/blv/en/home/tiere/tierschutz/wuerde-des-tieres.html#:~:text=The%20Swiss%20Animal%20Welfare%20Act,be%20respected%20when%20handling%20it.
Franklin, Julian H. *Animal Rights and Moral Philosophy.* New York: Columbia University Press, 2005.
Garner, Robert. *A Theory of Justice for Animals: Animal Rights in a Nonideal World.* Oxford: Oxford University Press, 2013.
Goodall, Jane. "Chimpanzees: Bridging the Gap." In *The Great Ape Project: Equality Beyond Humanity*, edited by Paola Cavalieri and Peter Singer, 10–18. New York: St. Martin's Press, 1993.
Herzberg, Guntolf. "Gerechtigkeit für Tiere." In *Tiere ohne Rechte?* edited by Jan C. Joerden, and Bodo Busch, 29–39. Berlin: Springer, 1999.
Jaeckle, Peter. *Amazing Pigs – Mysterious Boar.* Culver City, CA: United Seabears Corporation, 2011.
Marino, Lori, and Christina M. Colvin. "Thinking Pigs: A Comparative Review of Cognition, Emotion, and Personality in *Sus domesticus.*" *International Journal of Comparative Psychology* 28 Article 23589 (2015). DOI: https://doi.org/10.46867/ijcp.2015.28.00.04
Markell, Patchen. *Bound by Recognition.* Princeton, NJ: Princeton University Publishing, 2003.
Markell, Patchen. "Recognition and Redistribution." In *The Oxford Handbook of Political Theory*, edited by John S. Dryzek, Bonnie Honig, and Anne Phillips. Oxford: Oxford University Press, 2008.
Milburn, Josh. "Plant sentience and the case for ethical veganism." *Animal Sentience*, 33, no. 5 (2023): 1–2. DOI:https://doi.org/10.51291/2377-7478.1773.
Mill, John Stuart. *Utilitarianism.* London: Parker, Son, and Bourn, 1863.
Morris, Christoper. "Justice, Reasons and Moral Standing" In *Rational Commitment and Social Justice: Essays for Gregory Kavka*, edited by J Coleman, and Christopher Morris, 186–207. Cambridge: Cambridge University Press, 1998.
Nussbaum, Martha C. *Frontiers of Justice: Disability, Nationality, Species Membership.* Cambridge, MA: Harvard University Press, 2006.
Plumwood, Val. *Feminism and the Mastery of Nature.* London: Routledge, 1993.
Plumwood, Val. "Human Vulnerability and the Experience of Being Prey." *Quadrant* 39, no. 3 (March 1995): 29–34.
Plumwood, Val. *Environmental Culture: The Ecological Crisis of Reason.* London: Routledge, 2002.

Precht, Richard David. *Noahs Erbe – Vom Recht der Tiere und den Grenzen des Menschen*. Reinbek bei Hamburg: Rowohlt Taschenbuch Verlag, 2000.

Precht, Richard David. *Warum gibt es alles und nicht nichts?* Munich: Goldmann, 2011.

Regan, Tom. The Case for Animal Rights. Berkeley, CA: University of California Press, 1983.

Regan, Tom. *The Case for Animal Rights*. London: Routledge, 1984.

Regan, Tom. *The Case for Animal Rights*. Berkeley, CA: University of California Press: 2004.

Regan, Tom. *Defending Animal Rights*. Chicago: University of Illinois Press, 2006.

Regan, Tom. Interview on DVD *Live and Let Live*. Film directed by Marc Pierschel. Mindjazz Pictures, 2014.

Rigato, Emanuele, and Alessandro Minelli. "The great chain of being is still here." *Evolution: Education and Outreach* 6, no. 18 (June 2013): 1–6. DOI:https://doi.org/10.1186/1936-6434-6-18.

Schlosberg, David. "Ecological Justice for the Anthropocene." In *Political Animals and Animal Politics*, edited by Marcel Wissenburg, and David Schlosberg, 75–89. London: Palgrave Macmillan, 2014.

Schopenhauer, Arthur. *Die Welt als Wille und Vorstellung*. Paderborn: Voltmedia, 2005.

Schweitzer, Albert. *Die Ehrfurcht vor dem Leben*. Munich: C.H. Beck, 2008.

Sezgin, Hilal. *Artgerecht ist nur die Freiheit*. Munich: C.H. Beck, 2014.

Smith, Christie. "Articulating Ecological Injustices of Recognition." In *Political Animals and Animal Politics*, edited by Marcel Wissenburg and David Schlosberg, 61–74. London: Palgrave Macmillan, 2014.

Solar Foods. "Solein: Protein out of thin air." Published 2023. Accessed May 7, 2024. https://www.solein.com.

Taylor, Angus. "An Interview with Sue Donaldson and Will Kymlicka." *Between the Species* 17, no. 1 (2014): 140–165. DOI: https://doi.org/10.15368/bts.2014v17n1.7.

Terrill, Ethan C. "Plants, partial Moral Status, and practical Ethics." *Journal of Consciousness Studies* 28. No. 1–2 (2021): 184–209.

Terrill, Ethan C., and Walter Veit. "Better to be a Pig Dissatisfied than a Plant Satisfied." *Journal of Agricultural and Environmental Ethics* 37, no. 4 (2024):1–17. DOI: https://doi.org/10.1007/s10806-024-09922-4.

Universal Declaration of Human Rights. Approved and proclaimed by the General Assembly of the United Nations in 1948. Accessed November 1, 2015. un.org/depts/german/menschenrechte/aemr.pdf.

Volkelt, Johannes. *Arthur Schopenhauer*. Paderborn: Salzwasser Verlag, 2014.

Weil, Kari. *Thinking Animals: Why Animal Studies Now?* New York: Columbia University Press, 2012.

Wuketits, Franz M. "Hilft es den Tieren, wenn wir Menschen uns selbst als Tiere wahrnehmen?" *TIERethik* 1 (2015) 11–24.

CHAPTER 9

The Empirical Status Quo on the Right to Life (in Austria)

When we think about nonhuman animal rights in general and the right to life in particular, it is interesting to note that there are already certain nonhuman animal species that are granted a right which is somewhat similar to the human right to life, at least in some countries. Consider cows in India who have special protection due to religious causes, for example. In Austria, the killing of dogs and cats for the purpose of obtaining their meat or other products of their bodies is prohibited (Rechtsinformationssystem des Bundes (RIS) n.d.-a, § 6.2). Although the lives of dogs and cats in Austria are not threatened by human consumption of meat, many of them are legally euthanized.

The list of animal species whose lives are better protected (i.e. dogs and cats) in Austria is short. The Austrian Animal Protection Act nevertheless speaks (almost hypocritically) of a fundamental general "prohibition of killing." Elaborating further on this "prohibition of killing" (ibid., § 6), it then spells out the exceptions (see especially ibid., § 6.4.1, regarding the killing of nonhuman animals for the production of meat and animal products):

Prohibition of killing

§ 6 (1) It is prohibited to kill animals *without reasonable cause.*
(2) It is prohibited to kill dogs or cats for food or other products.

D. Schneeberger, *Envisioning a Better Future for Nonhuman Animals*, The Palgrave Macmillan Animal Ethics Series,
https://doi.org/10.1007/978-3-031-75332-9_9

(3) The killing of animals for the purpose of education, training and further training shall only be permitted at scientific institutions and only to the extent that it is indispensable for the intended purpose and cannot be replaced by alternative methods.

(4) Notwithstanding the prohibitions under subsections (1) and (2), the intended killing of vertebrates may only be carried out by veterinarians. This shall not apply

1. for the professional killing of farm animals and of animals for food (§ 32),
2. for the professional killing of animals within the framework of education, training and further training pursuant to paragraph 3,
3. for professional pest control,
4. in cases where rapid killing is absolutely necessary to spare the animal irremediable suffering.

(Rechtsinformationssystem des Bundes (RIS) n.d.-a, § 6; my translation and emphasis)

The title "prohibition of killing" suggests the existence of such a prohibition, and then exceptions are laid down so that a "right to life" (which is not of the same kind as the right to life of humans and does not imply the same permissions and prohibitions) is only granted to cats and dogs. For example, (seriously) ill cats and dogs may be euthanized by veterinarians through an injection, but this is not (yet) the case with (seriously) ill humans, at least in Austria. As another example, a legal difference is also made if one (unintentionally) runs over a human being or a nonhuman animal such as a cat, a deer, a dog, or a bee on a street. Regardless of this difference in criminal law, a duty to render assistance also applies to the injury of nonhuman animals, which, for various reasons, is not sufficiently fulfilled in practice in many cases: "§ 9. Anyone who has visibly injured an animal or put it in danger shall, insofar as this is reasonable for him, render the necessary assistance to the animal or, if this is not possible, arrange for such assistance to be provided" (ibid., § 9).

Although only dogs and cats are explicitly protected from being killed (for the production of animal products) in Austria, the Austrian animal protection law also provides certain far-reaching legal protection for other animal species. The ban on killing animals without "reasonable" cause extends not only to mammals or vertebrates, but in fact to invertebrates such as insects, snails and worms, too, for example: "According to § 6.1.

Austrian Animal Protection Act, the life of animals [...] is a protected legal interest and there must be a justification ("reasonable cause") for any killing of animals; according to the Austrian Animal Protection Law, this condition applies to all, including invertebrates" (Tierschutz Ombudsstelle n.d.). This prohibition thus requires that nonhuman animal life is respected and spared and protected if no reasonable reason can be given for its destruction. But which reasons are "reasonable," which are not, and who (which impartial, or even not speciesist, judge) determines that? The term "reasonable cause" was probably not unintentionally left undefined in the legal text; this vagueness offers the animal industry sufficient leeway for its daily business.

A practice that is also currently common in egg production in Austria (and also (still) within supposedly better organic farming) is the killing of male chicks of laying hen breeds. Raising these chicks and later using them for meat is considered economically unviable, as chicken breeds tend to be extremely "specialized," depending on the intended use. Chickens of laying breeds produce too little meat too slowly to actually be of economically interest (i.e. profitable) for meat production. Therefore, in egg production, about 50 per cent of the freshly hatched chicks are usually killed, either by shredding, gassing or suffocation. This practice not only leaves (also) the (vegetarian) consumer with an uneasy feeling, it can also be regarded as not according with the law, because purely economic reasons are—though this may come as a surprise—not "reasonable" causes for killing: "Economic reasons alone, such as saving labor, time and costs or the pursuit of profit maximization are not suitable to fulfil the requirements for the existence of a "reasonable cause"" (ibid., 2). The Austrian food brand "Ja! Natürlich" has acknowledged the discomfort many consumers feel about the issue of killing male chicks and started a project in which "dual-purpose" chickens are bred and the male chicks are raised for 90 days and then slaughtered for their meat (Ja! Natürlich n.d.). The brand "Tonis Freilandeier" (Freilandeier = "free-range eggs") has also tried to soothe the consciences of consumers with a similar campaign, and thus prevent a possible decline in egg sales (Toni's Freilandeier n.d.).

Germany has been discussing considerable legal steps towards eliminating the killing of male chicks of laying hen breeds. In the summer of 2015, a motion by the Federal Council of Germany (*Bundesrat*) was submitted to the Bundestag concerning a ban on the killing of male chicks after hatching: "According to the draft, the killing of animals for purely economic reasons is to be prohibited in the Animal Protection Act. A transitional period until mid-2017 should enable farms to adapt to this" (*Zeit Online* 2015; my translation). Through a procedure that recognizes the sex of the future chick in the egg, only female embryos of the laying hen breeds are to be hatched. "'My goal is for chick shredding to stop in 2017,' said Federal Minister of Agriculture Christian Schmidt (CSU). [...] In the future, the sex is to be determined after just three days in the incubated egg. The eggs can then be sorted out accordingly" (ibid.). The law was finally enacted in January 2022 (Bundesministerium für Ernährung und Landwirtschaft 2023). Breeders now determine the sex of the chicks before they hatch and kill the male embryos before hatching. Chicks normally hatch after 21 days in the egg (*Zeit Online* 2022). Determining the sex of a chick is however only possible 9–14 days after the egg has been laid (ibid.) and the chick embryos in the egg might be able to feel pain after just 7 days (ibid.).

It is not only chicks that are killed before they have had a chance to grow up. Generally, farmed animals produced for their meat are killed in infantile or adolescent stages, as their meat is then still tender then (cf. e.g. Grabolle (2012, 16f) for an overview of the usual slaughter age of the various species). Laying hens lay an egg (in their first and usually only year of life) almost every day—the original, natural laying cycle of the ancestors of the highly bred "performance laying hens" is 12 eggs per year (cf. e.g. Bruers 2015). Eggs laid by modern laying hens average 60 g; those of their non-highly bred natural predecessors weighed 30 g (ibid.). The laying performance of modern laying hens is comparable to human women menstruating daily and losing twice as much menstrual blood (ibid.).

More than 93 per cent of laying hens in the US are currently still kept in conventional cages (*Daily News* 2015). This is expected to change, as McDonalds USA and McDonalds Canada are phasing out the use of eggs from caged hens, with a final conversion deadline of 2025): "On Tuesday came what might be the biggest victory yet: McDonald's announced that it will stop using eggs from caged hens in the U.S. and Canada" (ibid.). An immediate change would not be possible, were McDonald's egg consumption to remain the same; the supplier companies will therefore have to adapt to the wishes of this major customer and change their production facilities if they want to continue to supply McDonald's in the future (ibid.).

Laying hens are usually slaughtered after one laying period, i.e. after 10–15 months, when their laying performance declines. Since the meat of chickens who are one year or older is deemed to be no longer suitable as roast chicken, the former laying chickens end up as soup chickens (for example). Chickens in meat production are bred to put on an very large amount of mass in a short time (especially on the breast). They are sometimes slaughtered after only four weeks (Hühner Haltung n.d.). In human terms, such an early killing is equivalent to killing a one-and-a-half-year-old infant that weighs 30 kg at slaughter (Bruers 2015). Pigs are slaughtered after six months, which is equivalent to a two-and-a-half-year-old human child weighing 70 kg (ibid.). Cows raised for their meat, who have been bred to gain weight rapidly, are killed (again in human terms) as six-year-olds with a body weight of 50 kg (ibid.).

Similar issues are problematic in turkey farming: "The bones and cartilage of the turkey legs cannot keep up with the rapid growth of the rest of the body, so in the last third of fattening the birds' limbs are often deformed. Breeding has increased the amount of breast meat by a quarter to 40 per cent of body weight since 1970, and the birds often cannot even stand properly at the end of the fattening period" (Grabolle 2012, 49); note, too that that fattening period is extremely short and intense, at 140 days (ibid., 16).

How nonhuman animals suffer and are kept and killed in the animal industry illustrates how human society currently does not weigh the value of these beings' lives higher than their culinary habits. Currently, the human majority is of the opinion that it is ethically permissible to kill them for culinary interests. One might hope and speculate that the human morality concerning a nonhuman animal right to life might be expanding to include more nonhuman animal species. There are three main reasons why many humans are more likely to support better protection of certain nonhuman animals' lives. First, the lives of beings that belong to species with which humans tend to have friendly or familial relationships of closeness tend to be better protected (e.g. in Western societies often dogs, cats, and partly horses). Second, the lives of those that humans tend to see as persons with individual personalities tend to be valued more (e.g. great apes, dolphins). Third, protection of individual beings due to species protection is intuitively understood, and advocated for by many humans (e.g. great apes, tigers, lions). As mentioned and discussed below, species protection can foster the protection of individual lives as well.

The first reason why certain nonhuman animals are more likely to be granted a right to life is based, on the one hand, on the fact that humans naturally want to protect beloved members of their narrow social circle and, on the other hand, on the fact that they have learned through their interactions with individuals of these species that they are also subjects and personalities whose weal and woe is important. A recognition of their uniqueness, their subjectivity, and individual personality could thus foster better protection for nonhuman animals.

It is not only cats and dogs that might be closer to a right to life; horses are increasingly treated better (at least in Western countries) for similar reasons. If they are kept as leisure or sport horses, they are more and more seen as family members and friends. In Austria, morally (if not yet legally) they already—at least in part—assume a similar status as companion animals like dogs or cats, which implies a greater consideration of their interests (including the interest in their own (continued) life). The consumption of horse meat is not very widespread in Austria, for example. US-Americans might have a similar moral intuition/culture when it comes to eating horses. According to a national poll conducted in 2022 by the American Society for the Prevention of Cruelty to Animals (ASPCA), 83 per cent of Americans oppose the slaughtering of horses (2022). The reality horses face, is however, often still grim, even in the US. In a report on horse transports from the US to a slaughterhouse in Mexico, the US animal welfare organization Animals' Angels states:

> Still many American horses are transported to the plant in Juarez, which is not appoved [sic!] by the European Union because of the use of the "Puntilla Knife method".
>
> Horses are still transported on inhumane open trailers, which offer the animals not enough head room and no protection from the elements. In addition, the risk of injuries is high due to the metal piping across the top of the trailer.
>
> The 800 mile long transport from the export pens to the plant in Jerez takes 18 hours—a journey way too long for these already weakened animals. (Animals' Angels Investigations and Advocacy 2015)
>
> Note that "*puntilla*," referred to above, is
>
> a traditional slaughter method in which a knife is plunged into the back of the neck to sever the spinal cord. The aim is to produce immediate collapse of the animal. Puntilla is not condoned as a stunning method by the World Animal Health Organisation (OIE) because there is concern that the animal could be conscious during and after the neck stab. Nonetheless, it is still used in some developing countries. [...] Some nerve pathways are often functional after the neck stab and therefore it is highly likely that the animals

> remain conscious in at least some modalities for the next part of the slaughter procedure. (Limon et al. 2012, 3)

Another group of animal species that are closer to a right to life than other nonhuman animals are those in which humans are most likely to recognize nonhuman animal persons (apes, dolphins). Many species of whales are protected, not only because of increased recognition of their status as persons, but primarily due to a risk of species extinction. While the species protection efforts that we currently see tend to be intuitively supported partly due to anthropocentric values concerning humans valuing a biodiverse fauna and natural environment, they nevertheless might prove helpful when it comes to establishing a nonhuman animal right to life.

Species protection of whales is, for example, enshrined in the "International Convention for the Regulation of Whaling," which was signed in 1946, and is supposed to be equivalent to a ban on whaling with certain exceptions (Audiovisual Library of International Law n.d.). Since 1986, there have been special permits for the capture of whales for scientific and traditional purposes. Japan, Iceland, and Norway have made use of them. The number of whales killed has been rising for years, which is seen by environmentalists as an abuse of the instrument of scientific permits, as well as evidence that the Convention is still relatively "toothless" (ibid.). Today, Iceland and Norway no longer recognize the ban, engaging in open commercial whaling. Greenland is permitted to whale for traditional purposes. To protect whales from being killed, activists are present at sea trying to prevent their killing:

> In its now banned fishing program, Japan wanted to kill 900 minke whales per year in the Antarctic. However, the fishing fleet, which devours an estimated 15 million euros of taxpayers' money, was only able to kill 251 whales last winter due to interference from "Sea Shepard". Japan's representative at the Whaling Commission, Joji Morishita, had compared opponents of whaling to "eco-imperialists" in November. Japanese citizens would not often eat whale meat, but did not want to have their right to do so banned by foreign countries, Morishita explained in Tokyo. (*Die Tageszeitung* (*taz.*) 2014)

Many humans intuitively want to protect whales. Stranded whales tend to be returned to the sea when possible, which is an example of humanity's altruistic impulses which also work across species barriers.

The brutal mass slaughter of dolphins that is common in Japan is criticized by the Western world. Particularly infamous in this regard is the whaling town of Taji, where dolphins are hunted every year between September and March. The dolphins are herded together in a bay, and the most beautiful specimens are sorted and then sold to dolphinariums worldwide; the remaining ones are brutally killed:

> For this purpose, the animals are stabbed with long spears and hooks from the boat. The badly injured, flailing and screaming animals are hoisted into the boats where they are further worked on with knives [...]. There can be no doubt whatsoever that this method of hunting is highly damaging to the animals in every aspect. Surviving animals are traumatized by what they have experienced and by their isolation from their school. (Benz-Schwarzburg 2012, 357)

Despite the outrage over this practice, Japan is determined to continue this bloody "tradition." There are particular economic reasons for the desire to sustain this, since the sale of certain particularly "beautiful" specimens to dolphinariums is a lucrative business (*Die Tageszeitung* (*taz.*) 2014). The fishermen also have a continuing interest in the slaughter of the dolphins, as they feed on fishes and are thus seen as competition by the fishermen (ibid.).

The protection of the lives of nonhuman animals belonging to endangered species is currently intuitively understood by many humans, and legally implemented. At some point, species protection could evolve into intuitive protection of the individual animal. Thus, the status as an individual belonging to an endangered species can co-establish a right to life of specific individuals (cf. e.g. Benz-Schwarzburg (2012, 333): "In the case of the Great Ape Project, the stridency of the animal rights demands raised was particularly understandable against the background of the extreme threat to great apes in the wild and the controversial captivity of these animals in zoos as well as their use in animal experimentation research").

Although lions are not endangered as a species, conservation efforts are necessary to keep their populations stable (World Wide Fund for Nature (WWF) n.d.). Humans might admire them "as a symbol of power, strength, and courage" (ibid.). Protecting those majestic beings—next to concerns to reduce unnecessary animal suffering—might also be one aspect of the outrage over the infamous killing of the lion "Cecil" in Zimbabwe:

> The killing of the 13-year-old lion named Cecil in Zimbabwe by an American hunter, 55, has caused a storm of outrage on the internet. At the beginning of July, Walter J. Palmer, a dentist living in the US state of Minnesota, tied a dead animal to a car, the chairman of the environmental organization Zimbabwe Conservation Task Force (ZCTF), Johnny Rodrigues, announced. Together with another hunter, he lured the feline predator out of the Hwange National Park. Hunting is legal outside the park. Palmer then shot at Cecil with a bow and arrow, but did not kill the animal. Only after 40 hours the hunters discovered and fatally shot the lion. In addition, they had tried in vain to destroy the lion's collar, which was equipped with a GPS transmitter and had been put on by researchers from the Oxford University-supported Hwange Lion Research. In the national park in the west of the country, Cecil was known for his black mane and was often sighted. [...] Around 250,000 people signed a petition addressed to Zimbabwe's President Robert Mugabe to stop permits to hunt endangered animals after Lion Cecil's death. The killing of Cecil was also criticized in the USA. Democratic Congresswoman Betty McCollum called for an investigation by the US Attorney General's Office and the Nature Conservancy. These should investigate a possible violation of laws, she said. (*Welt* 2015)

Before I move on to discussing a nonhuman right to freedom, I will first present an excursus on Norbert Hoerster's work *Haben Tiere eine Würde?* (*Do Animals Have Dignity?*). I return to normative ethics and critically discuss Hoerster's arguments against granting nonhuman animals a right to life. He contends that unlike humans, nonhuman animals lack a significant interest in survival, which he believes justifies their use for meat consumption. Contrary to Hoerster's stance and as has become evident in prior chapters of this book, I argue that nonhuman animals exhibit a strong interest in their survival. Hoerster further contends that nonhuman animals lack self-awareness and the capacity to benefit from rights due to their purported inability to relate to the future. While some might advise against "feeding the trolls" at this point, this is nevertheless an opportunity to practice ethical deliberation that I would like to engage in at this point.

References

American Society for the Prevention of Cruelty to Animals (ASPCA). "New Poll Confirms That Overwhelming Majority of Americans Oppose Horse Slaughter." Last modified February 9, 2022. Accessed February 10, 2023. aspca.org/news/new-poll-confirms-overwhelming-majority-americans-oppose-horse-slaughter.

Animals' Angels Investigations and Advocacy. "The Transport of American Horses to Mexican Slaughter Plants." Accessed August 30, 2015. animalsangels.org/images/stories/pdf/the%20transport%20of%20american%20horses%20to%20mexican%20slaughter%20plants.pdf.

Audiovisual Library of International Law. "International Convention for the Regulation of Whaling." Published n.d. Accessed February 10, 2023. https://legal.un.org/avl/ha/icrw/icrw.html.

Benz-Schwarzburg, Judith. *Verwandte im Geist – Fremde im Recht – Sozio-kognitive Fähigkeiten bei Tieren und ihre Relevanz für Tierethik und Tierschutz*. Erlangen: Harald Fischer, 2012.

Bruers, Stijn. "Stijn Bruers – All the Benefits of Veganism (IARC 2015)." Talk given at the International Animal Rights Conference in Esch, Luxembourg, 2015. Published September 12, 2015. youtube.com/watch?v=V6mfX7-Dbbc.

Bundesministerium für Ernährung und Landwirtschaft. "Verbot des Kükentotens." Last modified December 4, 2023. Accessed February 20, 2023. bmel.de/DE/themen/tiere/tierschutz/tierwohl-forschung-in-ovo.html.

Daily News. "Opinion: Macdonalds rattles the hen cage." Published September 12, 2015. Accessed October 20, 2015. nydailynews.com/opinion/peter-singer-mcdonald-rattles-hen-cage-article-1.2357723.

Grabolle, Andreas. *Kein Fleisch macht glücklich*. Munich: Goldmann, 2012.

Hühner Haltung. "Das produktive Huhn." Published n.d.. Accessed September 27, 2015. huehner-haltung.de/haltung/produktive-huhn.html

Ja! Natürlich. "Natürlich Bio-Eier vom Mooddorfer Haushuhn." Published n.d. Accessed September 24, 2015. janatuerlich.at/JaNatuerlich/Bio_Eier_vom_Moosdorfer_Haushuhn/Bio_Eier_vom_ Moosdorfer_Haushuhn/Content.aspx.

Limon, Georgina, Javier Guitian, and Neville G. Gregory. "A review of the humaneness of puntilla as a slaughter method." *Animal Welfare* 21, no. S2 (2012): 3–8. DOI: https://doi.org/10.7120/096272812X13353700593248.

Pasquet, Vivian. "Wenn jede Minute zur Qual wird." *Zeit Online*. Published January 15, 2015. Accessed November 27, 2015. zeit.de/2015/03/contergan-medizin-arzt-behinderung.

Rechtsinformationssystem des Bundes (RIS). "Bundesrecht konsolidiert: Gesamte Rechtsvorschrift für Verbotsgesetz 1947." Published n.d.-a. Accessed May 16, 2024. https://www.ris.bka.gv.at/GeltendeFassung.wxe?Abfrage=Bundesnormen&Gesetzesnummer=10000207.

Tageszeitung, Die (*taz*). "Jagd auf Delfine in Japann – Das Land des Gemetzels." Published September 16, 2014. Accessed February 15, 2015. taz.de/Jagd-auf-Delfine-in-Japan/!146071.

Tierschutz Ombudsstelle. "Der vernünftige Grund für die Tötung von Tieren." Published n.d. Accessed September 24, 2015. https://www.tierschutzombudsstelle.steiermark.at/cms/dokumente/11642258_50074656/2f4242b6/Tierärztekammer_Beitrag_Der%20vernünftige%20Grund%20für%20die%20Tötung%20von%20Tieren.pdf

Toni's Freilandeier. "Henne und Hahn." Published n.d. Accessed September 24, 2015. www.tonishennehahn.at.

Welt. "Cecils Jäger glaubt, alles richtig gemacht zu haben." Published July 29, 2015. Accessed October 6, 2015. welt.de/vermischtes/weltgeschehen/article144565352/Cecils-Jaeger-glaubt-alles-richtig-gemacht-zu-haben.html.

World Wide Fund for Nature (WWF). "Species: Lion – Facts." Published n.d. Accessed May 7, 2024. https://www.worldwildlife.org/species/lion%2D%2D19

Zeit Online. "Kükentöten wird ab 2022 verboten." Last modified January 20, 2021. Accessed November 7, 2022. https://www.zeit.de/politik/deutschland/2021-01/tierschutz-kueken-toeten-verbot-gesetzentwurf-bundeskabinett-agrarpolitik-landwirtschaft.

CHAPTER 10

Excursus: Norbert Hoerster, Nonhuman Animal Interests and the Right to Life

In his 2004 book *Haben Tiere eine Würde? (Do Animals Have Dignity?)*, Norbert Hoerster presents his views on animal ethics, which will be briefly discussed in this excursus, as in the course of this some counter-arguments against a nonhuman animal right to life can be addressed. In *Haben Tiere eine Würde?* Hoerster (2004, 73) negates that nonhuman animals should have a right to life. Hoerster (ibid., 72) argues, for example, that nonhuman animals, unlike humans, have no interest in survival that could be violated by their killing. That this view is motivated by Hoerster's interest in meat consumption is even suggested by Hoerster's (ibid., 71) own argumentation; against a ban on killing animals for meat consumption, he argues: "My view is that I myself and many other people like to eat meat, which speaks against a ban on this practice." Equivalently, a paedophile could argue for a right to have sex with children, since he and many other people enjoy having sex with children.

From the alleged non-existence of a nonhuman animal interest in survival, Hoerster (ibid., 73) concludes that the nonhuman animals should not have a right to life. Hoerster (ibid., 86) continues: "Insofar as animals have no interest in survival [...], they cannot feel any actual fear of death—apart from the fact that they also do not possess the mental capacities to infer their imminent end from their present situation." I argue the

D. Schneeberger, *Envisioning a Better Future for Nonhuman Animals*, The Palgrave Macmillan Animal Ethics Series,
https://doi.org/10.1007/978-3-031-75332-9_10

opposite, namely that nonhuman animals generally have a strong interest in their own survival and thus in not being killed. They reveal the existence of this urgent interest not only with their life-affirming actions (feeding, grooming, reproducing, etc.), but particularly forcefully (also) in situations where their own survival is threatened.

Hoerster (ibid., 72) moreover claims that certain nonhuman animals "such as pigs and cattle, but also dogs and cats" lack "self-awareness": "They do not have the capacity to experience themselves as individuals identical in the course of time with their own past and future" (ibid.). Hoerster (ibid., 100) also suggests that nonhuman animals could not benefit from an attribution of rights either, as they lack the "conscious future-relatedness of their experience." Hoerster (ibid.) continues: "In this respect, the individual animal cannot benefit at all—[...] like a human being—from an attribution and institutionalization of individual rights." Nevertheless, he maintains (ibid., 85) that experts (behavioural scientists, biologists) should educate laypeople about the abilities and needs that nonhuman animals have. This recommendation can certainly be agreed with.

Hoerster (ibid., 75) attempts to justify human meat-eating through utilitarian justifications that seem almost grotesque from an anti-speciesist perspective: "The human habit of eating meat from animals does not in fact reduce the quantity and thus the total value of animal life, but on the contrary increases it." From a position distanced enough from speciesism, this argument of Hoerster's could apply equally to the creation and killing of humans. One could imagine aliens populating and dominating the Earth and then also attempting to justify their (eventual) use of humans in such a way; for, one could argue as an alien in this scenario, if humans were not bred and killed, they would not exist in the first place. Similarly, a father who kills his daughter or son could argue that his actions are praiseworthy because he has (co-)created this being; until their death, his daughter or son would not have had a bad life, and thus he would have acted ethically commendable overall, since he has not diminished the total value of life, but on the contrary increased it. If someone wanted to breed humans in order to then kill them and consume their meat, then it would probably be ethically desirable to dissuade this someone from "producing" humans at all, or to stop him as soon as possible. In this case, it would be better if these lives were not produced, rather than producing these beings and then killing them. Hoerster (ibid., 75) illustrates his view with a quote from Leslie Stephen (1896, 236): "The pig itself has the greatest interest in the demand for bacon." No (healthy) living creature can seriously have "the greatest interest" in being killed and the subsequent processing of their dead body into meat products.

Life that is never conceived has no interests. For non-existent beings there is neither wellbeing nor suffering. Life that has been brought into existence, however, definitely has interests that can be violated; when these interests are disregarded, actions can clearly be morally reprehensible. If a morally bad and a morally neutral action are up for discussion, it is clearly better to choose the morally neutral one. Tatjana Višak (2015, 134) argues similarly when she sees existence and non-existence as incommensurable and (ibid.) further emphasizes: "Bringing a[n] [...] animal into existence that would otherwise not have existed does not benefit this [...] animal." Instead of making nonsensical comparisons of incommensurable states, Višak (2015, 134) suggests: "morality should be concerned with effects on sentient beings."

The unjustified act of violence (killing) as well as the harm inflicted on a nonhuman animal by the killing (with the exception of specific cases such as euthanasia) remain ethically problematic even if a nonhuman animal is kept well and killed as painlessly and in as stress-free a way as possible. This is precisely where animal rights approaches go beyond pathocentric animal welfare approaches. If I kill someone else—and every nonhuman animal is someone, and never "something"—I must have a sufficiently good reason to do so. The killing of an animal is justified in particular if it is necessary (see e.g. self-defence). In the situation in which Western humans finds themselves, one cannot speak of a necessity to kill nonhuman animals for the consumption of their meat and bodily products—except in extremely rare and exceptional cases.

In the next chapter, I first present reflections on the right to freedom. I explore diverse meanings and implications of freedom for both humans and nonhuman animals. While some suggest that nonhuman animals lack an intrinsic interest in freedom (Cochrane), this view is critiqued as ableist and reflecting Western linguistic biases. Discussions on interspecies communication emphasize listening to and respecting nonhuman animals' expressions, moving from speaking for nonhuman animals to theorizing from and as (non)human animals. Ultimately, the chapter advocates for a nuanced understanding of freedom that recognizes diverse species' needs and capacities, promoting ethical standards based on respect, responsiveness, and enabling individual agency. I then discuss the distinction between abolitionism and extinctionism and questions regarding the ethics of keeping "pets." In this part, I provide reflections on intriguing questions such as: Should I let my cat roam free even if she kills mice and birds on her hunting trips?

References

Hoerster, Norbert. *Haben Tiere eine Würde?* Grundfragen der Tierethik. Munich: Beck, 2004.

Stephen, Leslie. *Social Rights and Duties: Addresses to Ethical Societies. Volume I*. London: Swan Sonnenschein & Company, 1896.

Višak, Tatjana. "Do Utilitarians Need to Accept the Replaceability Argument?" In *The Ethics of* Killing Animals, edited by Tatjana Višak and Robert Garner, 117–135. Oxford: Oxford University Press, 2015.

CHAPTER 11

Specific Rights in Nonhuman Animal Rights Declarations (Part 1)

11.1 Article 3: Right to Freedom

> Freedom […]. [O]nly those from whom it has been taken have any idea what it means. (Wachowski et al. 2012)

The term "freedom" has a broad range of meanings from "total absence of restraint to merely a sense of not being unduly hampered or frustrated" (*Merriam-Webster* online dictionary n.d.-b). Freedom is intuitively valued, since "being free allows animals to pursue those things that are in their own interests" (Browning and Veit 2021, 1). Humans, for example, often rate the wellbeing of nonhuman animals kept in zoos less favourably than the wellbeing of nonhuman animals living in the wild, since they might consider the former to be "bored and sad" (ibid.). In line with this intuition shared by many, "Donaldson and Kymlicka condemn capturing animals and putting them in zoos, even in the most progressive zoos, as 'a violation of their basic individual rights, and a violation of their rights as members of sovereign communities'" (Keulartz 2021, 119). Castelló (2022, 145) suggests that Nussbaum seems to be in agreement when she states that "depriving animals of their differential ways of living and existing causes "a kind of premature death, the death of a form of flourishing that has been judged to be worthy of respect." In contrast to this view,

D. Schneeberger, *Envisioning a Better Future for Nonhuman Animals*, The Palgrave Macmillan Animal Ethics Series,
https://doi.org/10.1007/978-3-031-75332-9_11

Browning and Veit (2021, 2) make the case that "captive animals, when well-cared for, need not necessarily suffer and can even do well in captivity—perhaps even better than in the wild." "Interests in control or autonomy could [...] be outweighed by the factors that reduce autonomy in the wild such as predation, disease, and time spend [sic!] on foraging" (Browning and Veit 2021, 10).

When it comes to a nonhuman animal right to freedom, the question arises if humans are allowed to keep nonhuman animals in captivity at all. Before I explore Gary Francione's extinctionist approach (cf. Francione and Garner 2010), which suggests that we should cease keeping nonhuman animals in captivity, and that the institution of keeping pets should be abolished, I would like to provide a detailed account of what freedom (also in dependent, "unfree" conditions) can mean.[1]

The saying "one's freedom ends where another's begins" illustrates that in our societies, human freedom, too, is never total (cf. Browning and Veit 2021, 2). It is limited by moral and legal boundaries. Ceteris paribus, I am, for example, not free to kill a human. If I nevertheless do kill a human, the state will take away my freedom and put me in prison (except if the killing were necessary for self-defence, for example). We can therefore state that human and nonhuman animals do not have total freedom. Or to put it differently: "Freedom is a continuum" (ibid., 1). Human and nonhuman animals also do not need total freedom of action to be able to thrive. Browning and Veit (ibid., 2) elaborate: "[T]here are a number of

[1] I would like to briefly mention the "five freedoms" at this point. I will not discuss them in greater detail, since they are embedded in a welfarist paradigm which might not be too helpful for the purposes of this book. In the 1980s, the concept of the "five freedoms" was developed by the British Farm Animal Welfare Council (FAWC) (National Archives 2012). The degree to which these five freedoms are present is used to measure "animal-friendly" husbandry. These five freedoms are:

1. Freedom from hunger and thirst, by ready access to water and a diet to maintain health and vigour.
2. Freedom from discomfort, by providing an appropriate environment.
3. Freedom from pain, injury and disease, by prevention or rapid diagnosis and treatment.
4. Freedom to express normal behaviour, by providing sufficient space, proper facilities and appropriate company of the animal's own kind.
5. Freedom from fear and distress, by ensuring conditions and treatment, which avoid mental suffering.

ways in which captive animals can be free and in which wild animals can be constrained," suggesting that we ought to talk about "freedoms from" and "freedoms to." Even if nonhuman animals kept in zoos might not be free to roam around as they please, they might be free from (severe) hunger, for example. Browning and Veit (2021, 5) point out that nonhuman animals living in the wild cannot roam around wherever they please: "[T]hey are instead confined to strict territories, with strong mutual boundary control from and against their neighbours." Browning and Veit (ibid., 6) then argue that "even within their own territory, animals rarely wander freely, instead repeatedly following delineated tracks between important resources, and often at the same times each day." Nonhuman animals who live in the wild often do not have certain kinds of freedoms at their disposal. Browning and Veit (ibid.) emphasize that "wild animals are frequently unable to meet many of their desires." They do, however, point out (2021, 10) that it might not be possible to keep all kinds of nonhuman animal species in captivity. Some (e.g. great apes, elephants, or whales) might have needs which cannot be provided for to a sufficient extent in captivity (ibid.).

All in all, Browning and Veit's reflections make it clear that freedom is not a "simple dichotomy, and not all types of freedom will be equally valuable for welfare" (ibid., 7).

We need freedom, or autonomy, to certain degrees. Generally, we can state that "[t]he ability to move about and choose freely is associated with pleasure and satisfaction, while constraint can cause frustration and suffering (ibid., 8). Cochrane (2012) argues that nonhuman animals should not have a right to freedom since they do not need it. In his point of view, nonhuman animals lack an intrinsic interest in having ultimate control over their lives. For Cochrane (ibid., 206), nonhuman animals are not autonomous: "For autonomy entails the capacity to reflect on and choose one's own life goals. Without such capacities for reflection, animals are locked into their ends and desires in a way that most adult humans are not." However, Cochrane's general assumption that nonhuman animals do not reflect on their "life goals" might be open to doubt, since it can hardly be proven. The degree of freedom to fulfil one's own needs and to act autonomously is from my point of view also of importance for many nonhuman animals when it comes to their happiness and satisfaction in life. Martha Nussbaum (2010, 536) elaborates: "We must ask ourselves in each individual case to what extent the respective being is able to set goals and plans and to plan its life. Insofar as this ability is present, it should be promoted [and] [...] this promotion requires [...] measures [...]: Plenty of freedom of movement and the opportunity to engage in a range of different activities." In this sense, one can assume a nonhuman animal interest in the greatest possible freedom (to perform fulfilling, rewarding activities in a certain degree of self-determination) and a corresponding right to this freedom, at least to an extent that is sufficient for a happy and good life.

Cochrane's argument that nonhuman animals do not need a right to freedom is criticized as being ableist (e.g. Castelló 2022, 41) and formed under the influence of our violent Western linguistic heritage:

> [I]t is here where, perhaps, the relationship between Western language and human sovereignty becomes more apparent: the animal as *alogon*,[2] handless and eyeless, cannot have, for instance, an intrinsic interest in having a right to liberty and self-determination [...] and the human as the one with the *logos*, the sovereign gaze, and with hands to grab the staff and sword is sovereign over those who are un-able. (Castelló 2022, 91)

Rights certainly should be based on needs, and human and nonhuman animals do have diverging needs for different kinds of freedom. "Freedom may also be an intrinsic good to animal welfare if animals can value their own freedom in a stronger sense—in conceptualizing and desiring it. This is a type of 'second-order valuing', to value [...] freedom itself. This certainly seems to be true in the human case, where humans strongly value their own freedom" (Browning and Veit 2021, 8). When thinking about a right to freedom, autonomy is a central concept we should consider.

Autonomy[3] can be defined as "the quality or state of being self-governing" (*Merriam-Webster* online dictionary n.d.-a) or as "the capacity to think, decide, and act freely and independently on the basis of such thought and decisions" (Gillon 1985). Both "autonomy of will, or the capacity to decide to do things on the basis of one's deliberations" (ibid., 1806) and "autonomy of action" (ibid.) come with needs which are

[2] "Heidegger thought that animals lack language (*alogon*)" (Castelló 2022, 90).

[3] For a comprehensive account of "inclusive autonomy" of domesticated nonhuman animals, see Côté-Boudreau (2016).

observable in many nonhuman beings. Both human and nonhuman animals have needs to make decisions and to move freely. Those needs are visible in many nonhuman beings; nonhuman animals, too, "have an interest in shaping their own lives" (Meijer and Bovenkerk 2021, 50). Many nonhuman animals have a need for "inquisitive exploration" (ibid., 51). If this need is restricted, individuals "will not acquire certain competences and their level of interaction and self-expression will be limited" (ibid.). The need for autonomy, however, is present in different shapes depending on the species (ibid., 53) and the individual developmental and health condition. Human adults might have capacities and needs to autonomy which human children do not yet exhibit.

Human children and nonhuman animals who live with humans are dependent on human adults to enable agency and autonomy and ensure a "responsible exercise of freedom" (Donaldson and Kymlicka 2014, 26). Some nonhuman animals can thrive even if they live in captivity if their needs for autonomy are enabled to a large enough extent in order to spare them too much frustration. In some situations in multispecies societies, a restriction of freedom can also be necessary. Comparably, nonhuman animals living with human guardians might need parental restrictions (in order to stop them from endangering themselves on roads, to prevent them from harming others in the multispecies society, or to make sure they remain in an enclosure). Such restrictions can be compatible with a just anti-speciesist multispecies society. Not providing needed restrictions can constitute neglect, both in the case of human children or nonhuman animals.[4]

Alongside autonomy, agency is a central term in the debate about a nonhuman animal right to freedom. There are various kinds and understandings of agency. Agency can be defined as "the capacity for self-willed action" (Irvine 2004). An important aspect of agency is that "by the action the subject exerts an influence on the world around her" (Meijer and Bovenkerk 2021, 54). In order "for an action to be more than mere behaviour, [...] a desire or a will should be behind the action" (ibid.). Agency is "the expression or manifestation of a subjective existence" (Blattner et al. 2020a, 4). It is the "the capability of a subject to influence

[4] Compare this example taken from VINE sanctuary: "Rooster Baldy goes to great lengths to stay out of the coop at night, but his wishes are overridden because he isn't predator-savvy" (Blattner et al. 2020a, 8).

the world in a way that expresses her desires and will" (ibid.). Meijer and Bovenkerk (2021, 54) remind us not to fall into the trap of outdated anthropocentric scala naturae thinking: "Dog agency should be understood as dog agency, not as lesser-than-human agency."

(Eco)feminist scholars such as Carol Adams (2010), Josephine Donovan (2006) and Lori Gruen (2015) have explored relational agency. They emphasize that "all of us are born into webs of relations and are dependent on others at several points in our lives, to different degrees" (Meijer and Bovenkerk 2021, 56). Donaldson and Kymlicka explain that "if we don't view dependency as intrinsically undignified, we will see the dog as a capable individual who knows what he wants and how to communicate in order to get it—as someone who has the potential for agency, preferences, and choice" (Donaldson and Kymlicka 2011, 83). Castelló (2022, 139) refers to Donna Haraway's thinking when he states that "one does not have agency, agency is dynamic and always in the making (or unmaking)." What I would like to call "enabled agency" plays an important role in just multispecies interactions and societies:

> Meijer, Donaldson and Kymlicka, and Donovan and Adams argue that we should be responsive to other animals and "listen to what they tell us" [...]. It is by being attentive to animals' way of existence and "expecting agency, looking for agency, and enabling agency" [...] that we will be able to notice other animals as the kind of decision-makers that they are. Crucially, this will entail examining how non-human animals initiate actions of their own volition and author their own decisions. (Castelló 2022, 13)

An example of enabled agency provided by Blattner and colleagues (Blattner et al. 2020a, 8) is a cow in a sanctuary who waited at a gate for someone to open it. A human noticed this, and opened the gate for the cow. Castelló (2022, 193f) offers his interpretation of this situation:

> It seems clear, however, that this cow was protesting, that she was saying to us humans: "I do not want to be in the fenced area, I want to be with the herd in the upper hill". This form of resistance is political because the action demands through a non-linguistic form of protesting liberty. I would go as far as to say that the cow's action should be read as demanding institutional reforms to be granted rights to liberty, freedom of movement and association, and political participation.

The ethical imperative to enable and ensure nonhuman animal agency in just multispecies communities with humans has also been discussed by

Will Kymlicka and Sue Donaldson who talk about micro- and macro-agency (2014): "Micro-agency refers to the scope for making personal decisions, which can for domesticated animals include deciding on what to eat, where to sleep, and who to play with in the park or garden" (Meijer and Bovenkerk 2021, 57). Macro-agency refers to an ability to co-create interspecies communities concerning "larger power relations, reified in laws, institutions, and political and social processes" (ibid.).

In order to enable agency in multispecies communities, we need interspecies communication: "[W]e cannot speak for animals any longer. Animals speak, and so we ought to listen and theorise *from* animals and *as* animals" (Castelló 2022, 26). We need to make better efforts to be able to comprehend nonhuman languages (Meijer 2019a) and we should be attentive to "what the animals are telling us—rather than to what other humans are telling us about them" (Donovan and Adams 2007, 4).

Being good parents or good care-takers requires being "response-able" (Driessen 2014, 99). This means that those who are taking care of others make serious efforts to understand the wishes of those they take care of, take their wishes and needs into account, and enable them to "self-govern" themselves to a certain extent. Scholars co-constituting the recent animal labour discourse (cf. e.g. Cochrane 2016; Porcher 2017; Shaw 2018; Blattner et al. 2020b) have developed helpful understandings of how nonhuman animal voices can be heard and taken into account. Blattner (2018) describes an assent model that is already applied in the human realm in cases where living beings cannot actively consent and need a proxy to protect them and their interests. Assent is possible where individuals can express their preferences verbally, behaviourally, or emotionally (Black et al. 2010). This model is already in use in the human domain, for example, with human children or mentally handicapped humans. Seeking assent and respecting dissent (ibid.) (if possible) is paramount, in order to achieve ethical standards in shaping relationships which are characterized by a stark power imbalance. Furthermore, when engaging in interspecies communication, using and understanding body language is of special importance (Blattner 2018; Donaldson and Kymlicka 2011). Castelló (2022, 197) discusses the example of Eva Meijer and her dog companion Olli. In their shared animal language, "they develop a vocabulary of words and "gestures and body movements." This process enables Olli's political agency, i.e. he is enabled to be a being who can author more his daily practices, rhythms, decisions, and so on." Meijer (2019b, 100) illustrates how they co-authored their shared experiences: "I asked him things and taught him things, and he asked me things and taught me."

As mentioned above, these reflections on how we could live together in just multispecies communities stands in contrast to extinctionist views in the animal rights debate. Animal ethics extinctionism is a radical and controversial position located at the edge of the animal ethics continuum (Francione and Garner 2010) and advocates for the extinction of domesticated animal species. The aim is to bring about a separation between humans and the rest of the fauna, so that eventually there will be no more domesticated nonhuman animals. The most prominent representative of the extinctionist approach is the US legal scholar Gary L. Francione, who argues as follows:

> Domesticated animals are dependent on us in all kinds of ways. It is up to us when and whether they eat, whether they have water, where and when they are allowed to relieve themselves, when they sleep, whether they can get exercise, etc. Unlike human children […] domesticated animals belong neither to the nonhuman world nor entirely to our world. They remain forever in an underworld of vulnerability where they are dependent on us for everything that matters to them. […] They do not belong in our world and should not remain attached to it regardless of how well we treat them. (Francione 2007)

So, if you want to detach yourself from (your own) speciesism as much as possible, you might at some point, like Francione, start to ethically question *any* keeping of nonhuman animals.[5]

[5] Francione is a dog guardian himself and sees the dogs he adopts from dog shelters as "nonhuman refugees." Cf. his explanations in this regard during a public discussion with David DeGrazia at Pennsylvania State University in Philadelphia on May 17, 2016). In this sense, he (and I'm sure many others) judge his pet guardianship as ethically justified and as ethically good). Gary Francione regularly publishes the kill list of a New York animal shelter on his Facebook account. See, for example, this post from his Facebook feed in August 2015:

> Kill lists at New York City's Animal 'Care' and Control for Monday, August 17th 2015: CATS & KITTENS: http://bit.ly/1y9NUZI
> DOGS & PUPPIES: http://bit.ly/1H0dYYQ
>
> As of now, there are 36 wonderful cats and 14 wonderful dogs who will be killed because ACC does everything possible to discourage potential adopters and rescue groups who are trying to get the animals out. This is morally wrong and we all know it. Remember that every animal adopted or fostered is one more life saved and that means everything to that one animal. You save them and they save you. They teach you about love. Please help them stay alive! Please share this post. If you do not live in the New York City area, then go to whatever shelter is close to you and adopt/foster one or more animals of any species. And remember: the animals we love are no different from the ones we eat. Go vegan.

Is (the institution of) keeping pets, which usually involves a certain deprivation of freedom and, in addition, often castration and a negative influence on the species-appropriate social and sexual life of a nonhuman animal, actually speciesist? Should it therefore be abolished? In general, I myself believe that nonhuman animals *can* lead good lives if they are living with humans. Take cats, for example. If there is something to eat in a house, a warm and dry place to stay and then some friendly attention and interaction, even cats who are initially strangers to a household will quickly voluntarily move into an interspecies flat-share with humans. With a cat flap, they have all the freedom to come and go as they please. However, a cat who has unlimited outdoor access runs the risk of being injured or killed in traffic. The cat guardian must decide whether they want to confine cats; being confined, however, is naturally repugnant to cats (at least, if they are aware of the experiences that await them outside). Allowing them to roam freely, the guardians accept the risk of the cat being run over or otherwise injured. Additionally, as carnivores with a strong hunting instinct, many other nonhuman (prey) animals might be endangered by a free-roaming cat. We thus have to weigh the cat's interest to roam free against the interests of the prey animals not to become victims of the cat against each other. If carnivorous predators live in multispecies communities, we are faced with what I would like to call here the "domesticated predator problem" (cf. also the "predation problem" when it comes to wild animal suffering (cf. e.g. Dorado 2015; Keulartz 2021)). Do we have to prevent cats who live with us from killing prey animals?

Cat guardians are faced with the issue that cats are obligate carnivors, biologically speaking (in contrast to dogs, who are omnivores like humans and can thus more easily thrive on a vegan diet). While some companies are working on developing cultivated cat food at the moment (e.g. Meatly, a UK-based cultivated meat company), such cat food is not yet available. We know that animal bodies do not necessarily need specific foods, but need the necessary nutrients. Feeding cats a vegan diet might then again currently also put the owner in a legal greyzone (even if they cat does well healthwise on such a diet), since such a diet might not be considered species-appropriate and might, in the eyes of the law, go against animal welfare jurisdiction.

In Will Kymlicka and Sue Donaldson's *Zoopolis*, the claim is made that people who keep cats that have outdoor access have a duty to prevent them from killing prey: "Companion cats are part of our community, and that means we limit their ability to use force on other animals where possible, just as we would our children" (2014, 332). Donaldson and Kymlicka state that (ibid., 331): "cats belonging to a mixed human-animal society have no right to food that involves killing other animals." Companion cats are, according to Donaldson and Kymlicka (ibid.), "citizens" and thus obliged to respect the right of other animals not to be killed by them (or their guardians are obliged to ensure this). If cats have temporary and locally unlimited outdoor access, it is, however, probably not reasonable to expect a guardian to constantly monitor them when they are outdoors. Sezgin (2014, 217) disagrees with Donaldson and Kymlicka, and states that we do not have to prevent our cats from killing other nonhuman animals (as long as it is not a companion animal living in an interspecies community with humans). Sezgin (ibid.) also believes that while we have special duties of care for our companion animals, we at the same time have no duty to curtail their behaviours that may endanger nonhuman animals living in the wild. We have "special obligations towards the cat as someone who belongs to or is close to our family or circle of friends. On the other hand, we are no more responsible for their behaviour […] than we are for that of a fox that also hunts mice, or a hawk that hunts birds." Nevertheless, Sezgin (2014, 286) thinks that protecting a wounded mouse from a cat is ethically good, or rather imperative: "What if the cat puts a […] non-fatally injured mouse at our feet? I think that we can (or […] should) intervene here because, among other things, duties to help arise from the direct encounter with an animal in distress (similar to those towards a human being), especially since […] the aggressor does not need this animal as food."

Although companion animals are entitled to special protection and a certain level of care from their human guardians, and living as a "pet" thus comes with certain advantages, living in interspecies communities with humans entails certain costs for them. Living in a human household with access to the outdoors costs a cat their fertility without being aware of it, according to the law in Austria: "If cats are kept with regular access to the outdoors, they must be neutered by a veterinarian." (Rechtsinformationssystem des Bundes (RIS) n.d.-a, annex 1, 2/10). Cf. Sezgin (2014, 293): "It is no coincidence that castration is the price the cat has to pay for entering civilization […]. For nature's carefree way of scattering untold numbers of living beings into the world with uncertain outcomes so that some survive is not compatible with our morality."

While there might be (speciesist?) naturalistic intuitions that we do not have to prevent cats living with us from killing mice—"it's only natural"—we do have strong intuitions that we need to prevent nonhuman animals living with us from endangering humans. In this sense, human guardians are responsible for ensuring that their companion dog does not harm humans. Horses—and sheeps (see quotation below)—who are kept by humans, for example, must also be kept in secure and escape-proof fences; a free-roaming horse, for example, can quickly become a great danger to themselves and, for example, to humans driving cars on roads:

> Most people see [...] sheeps as "farm animals," but of course they can also be "kept" and allowed to live for their own purposes. In doing so, they are subject to minimal restrictions on their freedom, insofar as their (large and varied) pasture is fenced, partly because they do not understand the dangers of road traffic. None of us can run everywhere, and this limitation is certainly in their own interest. (Sezgin 2014, 236)

Consequently, a nonhuman animal can only be kept where their wellbeing and the development of their species-appropriate abilities and needs are sufficiently possible, despite certain necessary restrictions on individual freedom, to ensure a good and happy life for all members of the community. Castelló (2022) argues that ideally, we should change our cities and villages, in order to enable nonhuman animal freedom to a greater extent (e.g. make it safe for a dog to roam around on their own without being endangered by car traffic).

For domesticated nonhuman animals who (want to or have to) live close to humans, living conditions should be created that are species-appropriate for them and enable them to fulfil their needs and teloi as much as possible. When humans start to be more attentive to the voices, wishes, and agency of nonhuman beings, new ways of relating could become possible in the multispecies community.

The next section sheds light on the right to safety and physical integrity for nonhuman animals.[6] As it is rather straightforward and intuitive that nonhuman animals should have a right to safety and physical integrity similarly to humans, this chapter does not offer expansive normative argumentation. It instead provides an empirical description of what nonhuman animals currently experience at the hands of humans. It illuminates the

[6] For those who would like to read about nonhuman animal physical and *psychological* integrity in depth, publications by Marc Bekoff (cf. e.g. 2007, 2013) and Virginia Morell (2014) might be insightful.

harsh realities which are commonplace in the animal industry and in animal experimentation. It makes the case that currently, nonhuman animals are undergoing torturous procedures and processes even though legal frameworks in Western nations seemingly oppose "unnecessary" animal suffering. The chapter navigates controversial issues like ritual slaughter, which might involve nonhuman animals being killed while fully conscious. While this section does not focus on ethical normative argumentation, it mentions various animal ethics approaches which can be employed to argue for the protection of nonhuman bodily integrity (e.g. Tom Regan's "harm principle" and "respect principle," Peter Singer's "principle of equal consideration of interests," and Martha Nussbaum's capability approach).

11.2 Article 4: Right to Safety/Physical Integrity; Prohibition of Torture

In order to argue for the protection of individual safety and physical integrity, as well as protection against torture, one can draw on various existing animal ethics efforts. Tom Regan's "respect principle" and "harm principle," for example, can provide helpful guidelines. The respect principle states that we must treat those beings that have inherent value only in such a way that this value is always respected. If they are harmed, their value is in general not (sufficiently) respected. Regan believes that we have a prima facie duty not to harm animals with inherent value (harm principle) (cf. Rowlands 2013, 4f).

Singer's "principle of equal consideration of interests" might point to similar results if applied here, since—in many cases—a balancing of interests might show that the human interest under discussion weighs less than the interest of a nonhuman animal not to suffer harm. Utilitarian approaches allow constellations in which the interests of one individual may be overridden in favour of the profit of many individuals, and in some cases should be overridden. This feature of such approaches is problematic because it disregards the (infinite) value and significance of the individual's own life to the individual itself. It is indeed necessary to balance interests fairly—thus possible scenarios arise in which one may harm other animals, for example, in the case of self-defence—but nonhuman animal rights, just like human animal rights, should only be able to be trumped in certain exceptional cases.

Nussbaum's capability approach also results in a fundamental protection of bodily integrity: "Within the framework of the capability approach, animals have a direct claim not to be violated in their bodily integrity by the use of force, abuse and other forms of harm" (2010, 531). However, if there are good reasons, the protection of bodily integrity can be overridden, for example, with regard to the castration of nonhuman animals living with humans (ibid., 532f). According to Donaldson and Kymlicka (2013, 324), this, however, only applies where irresponsible procreation cannot be prevented sufficiently well or only too inconveniently in another, less drastic and irreversible way. This is of course in line with the principle of non-maleficence (Beauchamp and Childress 2013).

A certain prohibition against unnecessarily torturing and harming nonhuman animals is a central point of many Western legislations. In Austria, for example, it is prohibited under the Animal Protection Act to unjustifiably inflict pain, suffering or harm on nonhuman animals (Rechtsinformationssystem des Bundes (RIS) n.d.-b, § 51). The Austrian animal welfare legislation is welfarist in the sense of a visible ideal to avoid "unnecessary" suffering. Kymlicka and Donaldson (2014, 126) make the important point that the conceptual framework of avoiding "unnecessary" suffering is "catastrophic" for nonhuman animals. After all, "virtually all human violence against animals is unnecessary in the strict sense" (ibid.). Moreover, this framework does not prevent, but legitimizes suffering: "There are animal welfare laws on the books whose ostensible purpose is to curb the worst excesses of exploitation, but these offer scant protection, and legitimate rather than limit exploitation" (Donaldson and Kymlicka 2014, 42).

In order to avoid "unnecessary" suffering, in Austria (as in other countries), nonhuman animals who are slaughtered must be effectively stunned before they are actually killed, so that they are no longer conscious when they are killed: "Slaughtering animals without stunning before the withdrawal of blood is prohibited" (Rechtsinformationssystem des Bundes (RIS) n.d.-b, § 32.3). However, in every slaughterhouse there are a certain percentage of nonhuman animals for whom stunning before killing does not work, for a variety of reasons.[7]

[7] In May 2013, an inquiry was made at federal government level with regard to this error rate in stunning before the actual killing in slaughterhouses: "The error rate in the slaughter of cattle is in some cases over nine per cent, the answer says. In the case of pigs, the false stunning rate with hand-operated electrical equipment is as high as 12.5 per cent" (*Der Spiegel* 2012).

There are various methods of stunning during slaughter, which is supposed to reduce animal suffering in the process; some methods are judged to be more painful and distressing than others. The stunning of pigs in carbon dioxide pits in particular has been heavily criticized by animal rights activists. Excluded from the requirement of stunning before slaughter are ritual slaughters where religious rules prohibit stunning before killing. Such ritual slaughtering exists in Islam and Judaism, amongst other examples. In Austria, "[r]itual slaughtering without prior stunning of the slaughtered animals may only be carried out if this is necessary due to mandatory religious commandments or prohibitions of a legally recognised religious community" (ibid., § 32.5). It can be assumed that nonhuman animals suffer more when a ritual slaughter method is used than when they are killed with stunning beforehand. On the other hand, there are findings (cf. Schulze et al. 1978) that show on the basis of EEG measurements, that "stunning" by a bolt shot might also associated with significant pain.

Although in the Western world it has largely been legally agreed that nonhuman animals should not suffer "unnecessarily," nonhuman animals are kept in the animal industry in poor housing conditions that do not (or do not sufficiently) meet their needs, which could be judged as torture, because such conditions cause permanent suffering and damage to the animals concerned (cf. e.g. the squeezing husbandry in chicken farming[8]). Pigs—whose sense of smell is three times more acute than that of dogs (Schwarz and Berger-Lenz 2002, 18)—are kept in stables that are often

[8] Cf. e.g.: The health of the animals kept is also at great risk, especially due to the common poor (but legally compliant) housing conditions. In broiler farming, 20 to 26 chickens are kept on one square metre at the end of the fattening period (after three to four weeks). This type of husbandry is called "squeeze husbandry." These are stocking densities that significantly exceed the insufficient space available in laying batteries (17 animals per square metre), which is already recognised as inimical to animal welfare (Grabolle 2012, 58).

> It is not only the constant lying on the floor reeking of faeces, but also the air, which is heavily polluted with ammonia, that irritates the animals' skin. Fluid can accumulate in the lungs and abdominal cavity, the liver becomes fatty and the immune defences are weakened. It is therefore not surprising that almost all broiler chickens are treated with antibiotics in the course of their short lives, sometimes with different preparations. In half of the cases investigated by the North Rhine- Westphalian Ministry of Agriculture, the antibiotics were administered for only one to two days, although medical treatment would require three to six days. This indicates that the antibiotics were only used as growth accelerators, which is prohibited. As long as the use of medicines in poultry is not centrally recorded, this is unlikely to change, because many benefit from the system: the medicines enable profitability despite torture breeding and torture conditions, the treated animals grow faster, veterinarians and the pharmaceutical industry earn from the drugs sold, and consumers can continue to enjoy cheap "chicken meat." (Grabolle 2012, 59f)

upsettingly noisome, even for humans, with only 0.7 m allotted per pig. This lawful, conventional husbandry causes damage and suffering to the pigs, as they are prevented from carrying out natural activities and behaviours. In such environments, a species-appropriate, good and happy life is not possible because the pigs do not have sufficient freedom of movement and they cannot exercise a number of important species-appropriate activities. Karl-Heinz Schirnhofer, an industry insider, reacted in an open letter to video recordings in Austrian slaughterhouses published by the Austrian animal rights organization VGT, Verein gegen Tierfabriken ("The Association Against Factory Farms") in 2015:

> When I look at the secretly recorded videos of the well-known slaughterhouses in Austria, I feel the same as most other Austrians. I, too, am dismayed. [...] But the truth is that slaughterhouses—especially pig slaughterhouses—do not throw animal welfare out of the window for reasons of greed, but in order just to be able to survive economically! [...] Every year 15-20% of slaughterhouses close down. That was the main reason why we sold our slaughterhouse in spring 2015. [...] I know from practice that the trade sells cheaper than it buys the goods. In the past, this was accepted in the case of sales. In the meantime, it has become the rule. Meat and sausage products have become one of the biggest loss leaders. Retailers lose millions a year and they put up with it. [...] I have been preaching for years that we need to go back to the Sunday roast, and that we should drastically reduce our meat consumption. (Schirnhofer 2015)

It seems that despite massive (domestic and EU) subsidies for the animal industry, animal welfare measures are "unaffordable." If even those who profit directly from this industry are advocating that we should "drastically reduce" our meat consumption, what does that tell us about this industry? It is undoubtedly also the considerable subsidies that keep the animal industry going.

The air in stables where farm animals are kept is sometimes so bad that humans only enter them with respiratory protection.[9] Nonhuman animals

[9] Cf. e.g. the following description in an internet forum:

> I know what it's like in turkey barns because I've helped out with the stalls myself and you can't have a squeamish mindset because it's hard and you can only go into the barn with a respirator otherwise you can't stand it and the turkeys are so heavy that they have deformed bones and can't walk any more that's unfortunately really the case they have no more room to walk when they're about to be stalled and enough animals die there too 5% loss is already planned there. And without antibiotics none would survive until slaughter [sic!]. (schaf-foren.org n.d.)

who are kept in the animal industry may never see sunlight and never breathe fresh air (except perhaps on the day they are driven from the barn into the transporter to the slaughterhouse).

Procedures and forcibly induced conditions which, if inflicted upon humans, would be called torture, however, are everyday occurrences not only on factory farms, but also in animal testing in particular. Critical voices argue that it also brings no benefit to the progress of human medicine, but instead harms its progress. According to Aysha Akhtar (2015, 407), who cites data from the US Food and Drug Administration, an astonishing 92 per cent of drugs that have successfully undergone animal testing—i.e. that proved safe in animal testing and showed sufficiently discernible intended effects—are withdrawn after initial human trials. A further 2 per cent of drugs are withdrawn in their first two years on the market (cf. also remarks made by Hiltrud Strasser, a member of *Ärzte gegen Tierversuche* (Doctors Against Animal Experiments), in one of her lectures (2015)). This means that, in the final analysis, only 2 per cent of preparations successfully tested in nonhuman animal experiments are used for humans. Even with these 2 per cent, further complications cannot be ruled out; one example of this was the thalidomide scandal.[10] The active substance contained in the drug had not been found to have any harmful effects in nonhuman animal experiments; however, the foetuses of pregnant women who took thalidomide are well-known to have suffered severe malformations. According to critics of the animal experimentation industry, there is insufficient comparability of the effects of substances on organisms across species to actually be able to provide scientific prognoses for an effect in a human organism (cf. again remarks from Hiltrud Strasser's aforementioned lecture). Even between species that are genetically relatively close to each other, there are major differences in the effect of a substance: "According to scientific studies, mice and rats only have similar results of slightly more than 50% when substances are tested for carcinogenic effects" (Ärzte gegen Tierversuche n.d.). Concerning drugs which are administered when humans suffer strokes, we know that "[a]bout 1,009 potential drug candidates for stroke have been tested in animal

[10] "Mrs. Schwartau, when were you born?" "On 20 June 1962, but don't ask me the time. My parents were so shocked when they saw me for the first time that they forgot the time." Christiane Schwartau is sitting in her living room in Finkenwerder, her grey hair cut short. Where arms normally begin, hands peek out of sewn-on blouse sleeves. Three fingers on the left, three fingers on the right. (Pasquet 2015)

models, of which 97 [...] proceeded to clinical trials. Systematic reviews show that *only two* drugs are considered safe and effective for stroke: aspirin and tissue plasminogen activator" (Langley 2009, 4; my emphasis).

On the one hand, substances thus pass through animal testing studies that are ineffective or harmful to humans; on the other hand, possible substances that are helpful to humans are filtered out in animal testing and are not researched further. For example, penicillin, which was discovered by chance by Alexander Fleming in a petri dish in 1928, is lethal to rodents: if it had thus undergone the usual series of tests on laboratory animals, it might have been discarded in the course of these and branded as toxic (Ärzte gegen Tierversuche n.d.). Asbestos is carcinogenic to humans, but well tolerated by rats and hamsters, for example (ibid.). Arsenic, lethal to humans, is well tolerated by sheeps (ibid.). Aysha Akhtar, speaking in a video about a conference on animal experimentation in 2015 (Oxford Centre for Animal Ethics n.d.) summarizes as follows: "Animal experiments are very poor predictors of what we are going to find in humans." She (ibid.) reports on a study published in the *British Medical Journal* in 2007 (Perel et al. 2007), comparing scientific evidence about six different treatments for five different diseases obtained from clinical trials in humans and nonhuman animals: "What they found was that the animal data matched the human data only half the time. In other words, the animal data was no more likely to predict whether those treatments would be safe and effective in humans than a toss of the coin" (ibid.).

In view of these results, the argument that human medicine benefits from nonhuman animal experiments is invalidated. In this argument, the enormous suffering inflicted on laboratory animals has not yet been considered. To illustrate the cruelty of the experiments in excerpts, consider some examples of animal experiments carried out in Germany in recent years: (Ärzte gegen Tierversuche n.d):

> In genetically manipulated mice, pain is induced in various ways: two of the three branches of the sciatic nerve are cut, or an irritating mineral oil is injected into a hind paw, or a hind leg is tied off with a tight rubber band for an hour. Pain perception tests are then performed. Among other things, the mice are placed on a 52° C hot plate. Finally, the mice are killed to examine the nerves.
>
> Newborn mice are subjected to cold anaesthesia. For this, a mouse is placed on ice cubes for five minutes. Then the mouse is slowly warmed up again. To determine what effects this procedure has later in life, the mice are subjected to various behavioural tests on anxiety and depression behaviour

> at the age of 2 to 3.5 months. Among other things, the forced swimming test is carried out. In this test, a mouse is placed in a pool of water with smooth walls from which it cannot escape. The time until it gives up and stops swimming is measured. Giving up is considered a sign of depression. At the age of 145 days, all mice are killed.
>
> White-tufted marmosets are infected with orthopoxviruses, which have a potential for bioterrorism. Almost all animals develop severe symptoms: respiratory problems, fatigue, lack of appetite, nasal discharge, redness and pustules on the skin of the face, abdomen and mucous membranes, bleeding from the skin, intestines, liver and bladder, swollen lymph nodes. Most animals die within two weeks or are killed when they are dying. The animals' organs are examined.
>
> Beagle dogs have three cheek teeth of the lower jaw extracted under anaesthesia. 12 weeks later, three titanium implants are placed in their place. After 8 weeks, pieces of gum are cut out in the area of the implants. After another 8 weeks, chronic gum damage is to be achieved by shaving off the top layer of the mucosa with a drill. The tissue defects are covered with self-dissolving collagen matrix obtained from pig tissue. After a healing period of 12 weeks, all dogs are killed to examine tissue samples.

These descriptions constitute only a part of the torture that takes place in animal testing laboratories, financed by the pharmaceutical industry and taxpayer money. It is not only the experiments themselves, but also the conditions in which the nonhuman animals are kept in laboratories that are damaging and cause (severe) suffering. These laboratory animals suffer in their cages because they cannot fulfil many of their needs: they know no sun, no fresh air, can hardly move and cannot occupy themselves meaningfully. They are lonely and/or stressed in social constellations that are unpleasant for them. They suffer from hunger, thirst and from the diseases, injuries and/or genetic impairments that humans inflict or have inflicted on them before they are finally "disposed of."

One group of nonhuman animals who survived torture are the 66 chimpanzees who were held captive for experimental purposes in a laboratory in Liberia by the New York Blood Center, a company that supplies blood to hospitals in the US. These 66 chimpanzees were released on isolated islands after the laboratory was closed down (The Humane Society of the United States 2015). The New York Blood Center began experimenting on chimpanzees in Liberia in 1975; some of these chimpanzees were captured from the wild and others were bought from private farms. Over 400 chimpanzees died or were killed during the course of the experiments. When the New York Blood Center ceased the experiments, the survivors were finally shipped to six mangrove islands in 2008. There, the chimpanzees have no natural access to water or food; they have to be isolated, since they are (partially) infected with hepatitis B. On these islands, they are unable to feed themselves because there are no natural resources, and the New York Blood Center originally assured that these chimpanzees would be cared for by them for life. However, in March 2015, the New York Blood Center ceased the supply of food and water it had been providing. Four days later, a charitable group of humans stepped in to take over supplying these chimpanzees with food and water, so that they would not die of thirst and starvation. Since then, the Humane Society has cared for the chimpanzees with the help of donations. A Humane Society representative herself witnessed a chimpanzee trying to exchange food for a favour: he held out a leg to her, as he had probably often had to do in the experimental laboratory (ibid.). This interaction is telling of the grim past of this being: "During morning feeding, [veterinarian Richard] Ssuna points to a grizzled ape who is missing an arm. The vet says the animal is 'essentially a victim of torture'" (euronews.green 2022).

We should thus all be aware that in our midst today, in Western society, a vast number of nonhuman animals suffer torture and pain inflicted by humans. Although we cannot immediately and directly help many affected beings, we can immediately and directly distance ourselves from the torture of nonhuman animals by humans by no longer supporting the animal (testing) industry as consumers and by distancing ourselves from it as much as possible and working towards institutional societal anti-speciesist change.

Having examined some of the harsh empirical realities in this world in this chapter, we are now (re)turning to more philosophical and legal debates. The following chapter explores the right to be recognised as a person/(legal) subject before the law. This chapter discusses the concept of "personhood" and its ethical implications concerning nonhuman animals. The discussion illuminates historical extensions of personhood to broader groups and efforts to extend this status to nonhuman animals (the "personhood extension project," as I would like to call it). I provide information about initiatives that have tried to achieve greater protection for nonhuman animals using this extensionist strategy such as the Great Ape Project, initiatives to achieve Whale Rights, or The Nonhuman Rights Project. The chapter also sheds light on theoretical and practical reasoning abilities of nonhuman animals, challenging traditional views about exclusively human cognitive capacities. Critics of the personhood extension project advocate for using other terms and legal statuses that acknowledge the moral worth of nonhuman animals without relying on the anthropocentric ableist personhood concept: alternatives that are proposed are, for example, Pablo P. Castelló's "animalhood" and Maneesha Deckha's "beingness."

References

Adams, Carol J. *The Sexual Politics of Meat.* New York: Bloomsbury, 2010.

Akhtar, A. "The Flaws and Human Harms of Animal Experimentation." *Cambridge Quarterly of Healthcare Ethics* 24, no. 4 (October 2015): 407–419. DOI: https://doi.org/10.1017/S0963180115000079.

Ärzte gegen Tierversuche. "Welche und wie viele Tiere leiden in Tierversuchen." Published n.d. Accessed April 10, 2024. aerzte-gegen-tierversuche.de/images/pdf/ausstellunga1.pdf.

Beauchamp, Tom L., and James F. Childress. *Principles of Biomedical Ethics.* 7th ed. New York: Oxford University Press, 2013.

Bekoff, Marc. *The Emotional Lives of Animals: A Leading Scientist Explores Animal Joy, Sorrow, and Empathy – and Why They Matter.* Novato, CA: New World Library, 2007.

Bekoff, Marc. *Why Dogs Hump and Bees Get Depressed: The Fascinating Science of Animal Intelligence, Emotions, Friendship, and Conservation.* Novato, CA: New World Library, 2013.

Black, Betty S., Peter V. Rabins, Jeremy Sugarman, and Jason H. Karlawish. "Seeking Assent and Respecting Dissent in Dementia Research." *The American Journal of Geriatric Psychiatry* 18, no. 1 (2010): 77–85. https://doi.org/10.1097/JGP.0b013e3181bd1de2.

Blattner, Charlotte E. "Animals, the Forgotten Proletariat." Published December 12, 2018. https://youtu.be/HRINOyJe-90.

Blattner, Charlotte E., Sue Donaldson, and Ryan Wilcox. "Animal Agency in Community: A Political Multispecies Ethnography of VINE Sanctuary." *Politics and Animals* 6 (2020a): 1–22.

Blattner, Charlotte E., Kendra Coulter, and Will Kymlicka, eds. *Animal Labour. A New Frontier of Interspecies Justice?* Oxford: Oxford University Press, 2020b.

Browning, Heather, and Walter Veit. "Freedom and Animal Welfare." *Animals* 11, no. 4 (2021): 1148. DOI: https://doi.org/10.3390/ani11041148.

Castelló, Pablo P. "The Language of Zoodemocracy: Contesting Human Sovereignty Over Animals." PhD thesis. Royal Holloway, University of London, 2022.

Cochrane, Alasdair. *Animal Rights Without Liberation*. New York: Columbia University Press, 2012.

Cochrane, Alasdair. "Labour Rights for Animals." In *The Political Turn in Animal Ethics*, edited by Robert Garner and Siobhan O'Sullivan, 15–32. London: Rowman and Littlefield, 2016.

Côté-Boudreau, Frederic. "Inclusive autonomy: A theory of freedom for everyone." PhD thesis, Queen's University, 2016.

Donaldson, Sue, and Will Kymlicka. *Zoopolis: A Political Theory of Animal Rights*. Oxford: Oxford University Press, 2011.

Donaldson, Sue, and Will Kymlicka. *Zoopolis: Eine politische Theorie der Tierrechte*. Translated by Joachim Schulte. Berlin: Suhrkamp, 2013.

Donaldson, Sue, and Will Kymlicka. "Unruly Beasts: Animal Citizens and the Threat of Tyranny." *Canadian Journal of Political Science / Revue Canadienne de Science Politique* 47, no. 1 (2014): 23–45.

Donovan, Josephine. "Feminism and the treatment of animals: From care to dialogue." *Signs: Journal of Women in Culture and Society* 31, no. 2 (2006): 305–329. DOI: https://doi.org/10.1086/491750.

Donovan, Josephine, and Carol J. Adams. *The Feminist Care Tradition in Animal Ethics: A Reader*. New York: Columbia University Press, 2007.

Dorado, Daniel. "Ethical interventions in the wild: An annotated Bibliography." *Relations: Beyond Anthropocentrism* 3, no. 2 (2015): 219–238. DOI: https://doi.org/10.7358/rela-2015-002-dora.

Driessen, Clemens. "Animal Deliberation." In *Political Animals and Animal Politics*, edited by Marcel Wissenburg and David Schlosberg, 90–104. London: Palgrave Macmillan, 2014.

euronews.green. "These abandoned apes have a dark story to tell about US testing." Published April 24, 2022. Accessed February 11, 2023. https://www.euronews.com/green/2022/04/21/these-abandoned-apes-have-a-dark-story-to-tell-about-us-animal-testing

Francione, Gary L. "Blog: Animal Rights and Domesticated Nonhumans." Last modified January 10, 2007. Accessed March 28, 2015. abolitionistapproach.com/animal-rights-and-domesticated-nonhumans/.

Francione, Gary L., and Robert Garner. *The Animal Rights Debate: Abolition or Regulation?* New York: Columbia University Press, 2010.

Gillon, R. "Autonomy and the principle of respect for autonomy." *British Medical Journal* (clinical research edition), 290(6484) (1985): 1806–1808. DOI: https://doi.org/10.1136/bmj.290.6484.1806.

Grabolle, Andreas. *Kein Fleisch macht glücklich*. Munich: Goldmann, 2012.

Gruen, Lori. *Entangled Empathy: An alternative ethic for our relationships with animals*. New York, NY: Lantern Books, 2015.

Humane Society of the United States, The. "New York Center Research Chimps Abandoned in Liberia." Published September 2015. Accessed November 4, 2015. humanesociety.org/news/magazines/2015/09-10/new-york-blood-center-research-chimps-abandoned-in-liberia.html.

Irvine, Leslie. *If You Tame Me: Understanding our connection with animals.* Philadelphia, PA: Temple University Press, 2004.

Keulartz, Jozef. "Should the Lion Eat Straw Like the Ox? Animal Ethics and the Predation Problem." In *The International Library of Environmental, Agricultural and Food Ethics Volume 33: Animals in Our Midst: The Challenges of Co-existing with Animals in the Anthropocene*, edited by Bernice Bovenkerk, and Jozef Keulartz, 99–122. Cham, Switzerland: Springer, 2021. DOI: https://doi.org/10.1007/978-3-030-63523-7_6.

Kymlicka, Will, and Sue Donaldson. "Animal Rights, Multiculturalism, and the Left." *Journal of Social Philosophy* 45, no. 1 (2014): 116–135. DOI: https://doi.org/10.1111/josp.12047.

Langley, Gill. (2009) "The validity of animal experiments in medical research." Last modified 2009. Last accessed 30 April 2024. https://www.wellbeingintlstudies-repository.org/cgi/viewcontent.cgi?article=1050&context=acwp_arte#:~:text=by%20Gill%20Langley&text=So%2C%20in%20medical%20research%2C%20experiments,lesser%20depending%20on%20the%20model, 1–6.

Meijer, Eva, and Bernice Bovenkerk, B. "Taking Animal Perspectives into Account in Animal Ethics." In *The International Library of Environmental, Agricultural and Food Ethics, Volume 33: Animals in Our Midst: The Challenges of Co-existing with Animals in the Anthropocene*, edited by Bernice Bovenkerk and Josef Keulartz, 49–64. Cham, Switzerland: Springer, 2021. DOI: https://doi.org/10.1007/978-3-030-63523-7_3

Meijer, Eva. *Animal Languages: The Secret Conversations of the Living World.* Translated by L. Watkinson. London: John Murray, 2019a.

Meijer, Eva. *When Animals Speak: Toward An Interspecies Democracy.* New York: New York University Press, 2019b.

Merriam-Webster online dictionary. "autonomy." Published n.d.-a. Accessed March 30, 2024. https://www.merriam-webster.com/dictionary/autonomy.

Merriam-Webster online dictionary. "freedom." Published n.d.-b. Accessed March 30, 2024. https://www.merriam-webster.com/dictionary/freedom.

Morell, Virginia. *Animal Wise: How We Know Animals Think and Feel.* New York: Crown, 2014.

National Archives, The. "Farm Animal Welfare Council: The Five Freedoms." Archived October 10, 2012. https://webarchive.nationalarchives.gov.uk/ukgwa/20121010012427/http://www.fawc.org.uk/freedoms.htm

Nussbaum, Martha C. *Die Grenzen der Gerechtigkeit: Behinderung, Nationalität und Spezieszuhörigkeit.* Translated by Robin Celikates and Eva Engels. Frankfurt am Main: Suhrkamp, 2010.

Oxford Cenre for Animal Ethics. "Ethics of Animal Experimentation." Published n.d. Accessed February 2, 2023. https://www.oxfordanimalethics.com/what-we-do/ethics-of-animal-experimentation/.

Pasquet, Vivian. "Wenn jede Minute zur Qual wird." *Zeit Online*. Published January 15, 2015. Accessed November 27, 2015. zeit.de/2015/03/contergan-medizin-arzt-behinderung.

Perel, Pablo, Ian Roberts, Emily Sena, Philipa Wheble, Catherine Briscoe, Peter Sandercock, Malcolm Macleod, et al. "Comparison of treatment effects between animal experiments and clinical trials: systematic review." *British Medical Journal* 334 (2007): 197–203. DOI: https://doi.org/10.1136/bmj.39048.407928.BE.

Porcher, Jocelyne. *The Ethics of Animal Labor: A Collaborative Utopia*. Cham, Switzerland: Palgrave Macmillan, 2017.

Rechtsinformationssystem des Bundes (RIS). "Bundesrecht konsolidiert: Gesamte Rechtsvorschrift für 2. Tierhaltungsverordnung." Published n.d.-a. Accessed May 20, 2024. https://ris.bka.gv.at/GeltendeFassung.wxe?Abfrage=Bundesnormen&Gesetzesnummer=20003860.

Rechtsinformationssystem des Bundes (RIS). "Bundesrecht konsolidiert: Gesamte Rechtsvorschrift für Tierschutzgesetz." Published n.d.-b. Accessed May 20, 2024. https://ris.bka.gv.at/GeltendeFassung.wxe?Abfrage=Bundesnormen&Gesetzesnummer=20003541.

Rowlands, Mark. "Animal Rights." In *The International Encyclopedia of Ethics*, edited by Hugh LaFollette, Hoboken NJ: Wiley-Blackwell, 2013. (Penultimate version. Accessed October 28, 2015. academia.edu/17280516/Animal_Rights.

schaf-foren.org: Fachforum für die Haltung und Zucht von Schafen und die Verarbeitung von Produkten der Schafhaltung. Published n.d. Accessed May 2, 2016. schaf-foren.de/forum/index.php?showtopic=17127&st=25.

Schirnhofer, Heinz. "Fisch + Fleisch: Schweineskandal – Stecken wir nicht alle im Boot?" Published November 21, 2015. Accessed November 25, 2015. https://www.fischundfleisch.com/styrias-best/schweineskandal-stecken-wir-nicht-alle-im-boot-12053.

Schulze, Wilhelm, H.S. Schultze-Petzold, A.S. Hazem, and R. Gross. "Versuche zur Objektivierung von Schmerz und Bewußtsein bei der konventionellen (Bolzenschußbetäubung) sowie religionsgesetzlichen ('Schächtschnitt') Schlachtung von Schaf und Kalb." *Deutsche Tierärztliche Wochenschrift* 85, no. 2 (1978): 62–66.

Schwarz, Peter, and Monika Berger-Lenz. *All About Mini Pigs*. Norderstedt: Books on Demand, 2002.

Sezgin, Hilal. *Artgerecht ist nur die Freiheit*. Munich: C.H. Beck, 2014.

Shaw, Rosemary. "A Case for Recognizing the Rights of Animals as Workers." *Journal of Animal Ethics* 8, no. 2 (2018): 182–198. DOI: https://doi.org/10.5406/janimalethics.8.2.0182.

Spiegel, Der. "Regierung rügt Tierquälerei in Schlachthöfen." Published June 21, 2012. Accessed November 10, 2015. spiegel.de/wissenschaft/natur/schlachthoefe-arbeiten-mit-hoher-fehlerquote-tiere-leiden-unnoetig-a-840156.html.

Strasser, Hiltrud. Lecture entitled "Can Animal Experiments in the Laboratory be Scientific?" University of Salzburg. November 26, 2015.

Wachowski, Lana, Lilly Wachowski, and Tom Tykwer, directors. *Cloud Atlas*. Cloud Atlas Productions, X Filme Creative Pool, and Anarchos Productions. 2012.

CHAPTER 12

Specific Rights in Nonhuman Animal Rights Declarations (Part 2)

12.1 Article 5: Right to Recognition as a Person/(Legal) Subject Before the Law

12.1.1 Are there Nonhuman Animals Who Are Persons? The Concept of "Person" and Its Ethical Relevance

The division into persons and non-persons is by no means insignificant from a legal and ethical point of view. A human person enjoys a certain legal protection which other sentient beings are denied. The dignity (= inherent value) of a human person is recognised. This tends to be denied to nonhuman animals who are in general not recognised as persons. Meanwhile, the concept of "person" is also viewed critically in the animal ethics discussion, as it is rooted in a speciesist-perfectionist outlook. Maneesha Deckha (2021) points out that the category of legal personhood is anthropocentric, and thus not helpful when it comes to the project of creating just multispecies societies. The personhood concept is a tool of discrimination, as it is also used within a species (although intuitively definitely justified in part) to divide human beings into groups that have different rights and obligations. From a legal point of view, the personhood concept is also flexible and stretchable, as it is not limited to human beings. Think, for example, of nonhuman entities called "legal persons" (e.g. associations, corporations, foundations, etc.).

D. Schneeberger, *Envisioning a Better Future for Nonhuman Animals*, The Palgrave Macmillan Animal Ethics Series,
https://doi.org/10.1007/978-3-031-75332-9_12

The interdisciplinary project to extend personhood to nonhuman animals has been taking place for many years already. Critical voices (such as Deckha) point out that those who work for this extension might have the best intentions, but might nevertheless be misguided. Castelló (2022a, 64) illustrates its limitations: "The day may come when dolphins, great apes, and elephants have fundamental rights, while racism, sexism, and ableism remain as alive as they are today, and many other animals continue to be slaughtered, contained, experimented upon and consumed "without any apparent bounds" and on "an unimaginable scale." Those critical of the personhood extension project instead suggest that we should come up with new terms for a new legal status for (nonhuman) animals (Castelló 2022a, 115). In his visions of a possible future "Zoolondopolis" Castelló (2022a, b) suggests the term "legal animalhood." Legal animalhood is "a shared legal status for humans and animals" that recognizes "certain fundamental legal rights for all animals, such as legal standing, a right to flourish, liberty, and not to be harmed by humans" (Castelló 2022a, 239). Castelló (ibid.) elaborates on how this term differs to Deckha's (2020) term "beingness":

> Maneesha Deckha has recently proposed the category of legal beingness. I propose legal animalhood instead because its purpose is substantially different to Deckha's, i.e., Deckha seeks to substitute personhood for another ideal legal category, while legal animalhood is partly a strategic category. Further, in my view, animalhood encapsulates best the idea that animals (as opposed to non-animal beings) should have a status other than personhood and property.

Deckha's term "beingness" is "directed at eliminating the property paradigm for animals" (Deckha 2021, 8). "Beingness" is anti-anthropocentric in the way that it is "not rooted in respecting animals out of sameness to humans and human benchmarks" (ibid., 10). The central features of beingness are embodiedness, relationality, and vulnerability (Deckha 2021). Deckha (ibid., 35) explains that "[b]eingness as a new legal subjectivity is meant to carry the same high level of protection for animals that personhood is intended to provide by the latter's proponents, but without personhood's anthropocentric limits. Deckha (ibid., 123) points out that "Germany and Austria have inaugurated fellow-beingness status for animals in their jurisdictions […]. [T]he concept was inaugurated to recognize dignity, humanity, compassion, and justice in animals, bestowing them with inherent value, dignity, and identity." Since we know how nonhuman animals are currently treated in these countries, we can see that, there at least, "anthropocentric interpretations of beingness have precluded any revolutionizing impact" (ibid.).

Whilst critics of the "personhood extension project" make a convincing case in arguing that it is anthropocentric and thus deficient and misguided, I nevertheless think it might have pragmatic appeal in the non-ideal world we inhabit. I would thus like to now shed light on the personhood extension project in this section.

In the course of history, the concept of "person" was steadily extended to an increasingly large group of humans. For example, human slaves of the nineteenth century were excluded from personhood; rather, they were categorized as possessions, holding a status roughly equivalent to that of nonhuman animals. Their owners therefore had the right to sell them or even (under certain circumstances) kill them. Today, most humans are granted personhood. However, the question is whether (some) nonhuman animals also fulfil the criteria to be persons who should be respected accordingly.

That the existence and recognition of nonhuman natural persons is possible, and that even non-animal natural entities can be categorized as persons with corresponding rights, is shown, for example, by the fact that in 2012 the Whanganui River in New Zealand was classified as a person (Tanasescu 2014, 161). This categorization was also accompanied by a right of action (by proxy) on the part of the river (ibid.). Furthermore, the river now has two guardians; one a representative of the British Crown, and one a member of the groups of indigenous humans living along the river (ibid.). However, a meaningful extension of the personhood concept need not extend to inanimate natural entities. If one takes a look around at the nonhuman fauna (with open eyes and an open heart), it is teeming with nonhuman animals who (could) meet all the criteria to be considered a person.

It has already been argued by Peter Singer (amongst several others) that there are nonhuman animals who fulfil the criteria that a being must fulfil in order to be considered a person. The terms "human being" and "person" are not synonymous (Singer 1994, 120). Singer (ibid., 155) takes up the person-non-person distinction and argues for the fact that certain nonhuman animals can or should be considered persons. Expanding the "person circle," similar to expanding the "moral circle" (Singer 2011), bears a hope for better protection of those who now also have become persons.[1,2]

According to Singer (1994, 120), nonhuman animal persons should be accorded a right to life. Richard David Precht (2000, 247) criticizes Singer's argumentation on two points in particular. First, he questions how one is to know exactly who is to be considered a person and who is not: "What about frogs, for example? You have to know a lot about the consciousness of frogs to be able to decide whether frogs have elementary self-awareness—a lot more than is really known at the moment." Secondly, Precht (ibid.) criticizes the unconditional coupling of the protection of life to personhood, which would have intuitively unacceptable consequences; that is, human non-persons would also fall outside the protection implied by the personhood status (cf. a possible consequence of the argument of species overlap). In his *Practical Ethics*, Singer refers to the view of the philosopher Michael Tooley: "In order to have a right to life, one must—at least at some point—have (had) the idea of a continuing existence" (Singer 1994, 133). If this claim is accepted, then there is apparently nothing preventing the justification of killing human beings who have never had an awareness of their continued existence. Thus it would follow that mentally severely impaired human beings who may never have had an awareness of their continued existence would not be granted a right to life.

Although the personhood concept has its weaknesses, it is a recognised and important part of Western legal systems. Since it shows some potential for animal ethics purposes, we will now discuss in detail why nonhuman animals can also be persons and whether an argument for the increased legal protection for nonhuman animal persons can be derived from this.

In general, caution is advisable when discussing whether certain beings exhibit certain characteristics or abilities. As long as one does not have sufficiently strong evidence, it may be wise not to jump to conclusions and to bear in mind that certain questions cannot yet be (fully) answered. Epistemic humility is strictly advised.

[1] The "personhood extension project" has parallels to the idea to expand the term "worker" to nonhuman animals (cf. the "labour-recognition-transformation thesis" (Blattner et al. 2020, 5; Eisen 2020, 139ff.)). (Cf. Sect. 17.2 of this book).

[2] Moral circle expansion could also be crucial to tackle substratism, the discrimination of beings due to the substrate they consist of (cf. e.g. Pauketat and Anthis 2022).

Although it is by no means true that humans are the only animals that can speak, for example, this false assumption has, for example, been haunting humanity's minds for centuries; little wonder, since it was the highly influential Aristotle who said: "man alone of the animals possesses speech" (1944, 1.1253a). Singer (1994, 151) argues against this claim: "One could argue that many animal species use language, though not ours." Nonhuman animals simply have different languages or communication systems. Meijer (2013, 46) explains that "[t]he question of what counts as language is also a political question. Studying human-animal language-games shows us that a narrow conception of language, as belonging solely to humans, is arbitrary and untenable, as well as biased."

The influential definition of the term "person" proposed by John Locke in 1690 in *An Essay on the Human Mind*, which also shaped Singer's concept of person (cf. Singer 1994, 120), is the starting point of the discussion of the concept of person presented here. Step by step, the following paragraphs discuss whether there are nonhuman animals that exhibit the respective essential qualities of persons demanded by Locke. In *An Essay on the Human Mind*, Locke (2000, 419–420; my emphasis) describes his concept of a "person" as follows:

> In my opinion, this word denotes *a thinking, understanding being that possesses reason and deliberation and can regard itself as itself. That is, it grasps itself as the same thing that thinks at different times and in different places.* This happens merely through *consciousness*, which is inseparable from thinking and, as it seems to me, belongs to its essence. For it is *impossible for someone to perceive without perceiving that he is doing it.* When we see, hear, smell, taste, feel, consider, or will something, we know that we are doing it.

Thus, according to Locke, a being is a person precisely when it has four essential qualities or capacities: A living being is a person when it is a rational, thinking being that is aware of itself and can conceive of itself as the same (thinking) being at different times and places.

Theoretical and Practical Reason

In order to answer the question of whether (some) nonhuman animals are rational, we must first clarify precisely what "rational" means. Reason as a concept has been divided into theoretical and practical reason at least since Kant. Wallace (2009, 3) describes theoretical reason along the following lines: on the one hand, one is theoretically reasonable when one reflects on past events and asks oneself why they happened as they did; on the other hand, one is also theoretically rational when one reflects on future events (ibid.). One is also theoretically rational when one reflects on one's values and adapts them if necessary (Wallace 2009, 2). Practical reason, on the other hand, regulates our will and actions through certain principles (Conradi and Stollreiter 2004, 351).

When one acts practically rationally, according to the general understanding of the term, one consciously takes actions that seem appropriate to achieve a goal. Practical reason therefore does not influence our values or attitudes, but directly influences our actions (Wallace 2009, 3). For example, it can be seen as practically reasonable to ration food if a limited ration of food is available for a certain period of time.

Brensing (2013, 120f) elaborates on the practically reasonable actions of the dolphin Kelly in a dolphinarium in Mississippi. The dolphins in this dolphinarium were trained to pick up objects and rubbish from the tank and trade them for dead fishes, i.e. food. Kelly proved to be particularly clever in that she tore pieces of paper into smaller pieces and then collected a fish with each piece. One day she brought a dead gull to her trainer. He rewarded her with several fishes. Kelly then began catching gulls by saving the last fishes of her regular feeding and using them as bait for the gulls, which she then traded for several fishes. These actions illustrate that Kelly was a clever "entrepreneur," impressively smart and practically rational.

Singer (1994, 154) describes an observation made by Jane Goodall of a chimpanzee performing a practically rational act while studying the behaviour of this primate species in Tanzania:

> One day, not long after the group had been fed, Figan spotted a banana that had been overlooked—but Goliath [an adult male who was above Figan in the group hierarchy] was sitting right under it. After no more than a quick glance from the fruit to Goliath, Figan moved away and sat on the other side of the tent so that he could no longer see the fruit. Fifteen minutes later, as Goliath stalked away, Figan climbed over without a moment's hesitation and picked up the fruit. Obviously he had sized up the situation: If he had climbed over to the fruit earlier, Goliath would have pretty much snatched it away from him. If he had stayed near the banana, he probably would have checked on it from time to time. Chimpanzees are very quick to notice and interpret the eye movements of their comrades, so Goliath might have seen the fruit himself.

Figan's reaction obviously presupposes certain cognitive capacities. Singer (1994, 155) sees this example as confirmation that Figan is "aware of himself as a distinct entity existing in time." Practical reason thus also exists in nonhuman fauna; in general, however, it is more difficult to verify whether nonhuman animals are also theoretically reasonable, since we currently do not have direct access to their streams of consciousness, i.e. to their perceptions, sensations, and thoughts.

Ability to Think

Anyone who studies nonhuman animals in depth will probably object to denying nonhuman animals the ability to think. Shoemaker (2008, 314) finds it alarming to claim that nonhuman animals cannot think, just as he rejects the claim that humans are not animals. Singer (1994, 150) points out that there are philosophers—and he cites Stuart Hampshire and Michael Leahy as examples—who believe that thought requires language (sufficiently similar to a human language). One may well doubt that (human) language is a necessary precondition for (nonhuman) thought. Perhaps there is more truth in the reverse: thinking precedes the development of language. Singer (1994, 152) notes critically that: "Hampshire and Leahy are arguing quite typically like many other philosophers who hold the same views: they are trying to philosophize from their chair in a field that requires research in the real world."

Self-Awareness

> Not only the *consciousness of other things*, i.e. the perception of the external world, but also self-consciousness *contains* [...] *a cognizing and a cognized*: otherwise it would not be consciousness. [...] For as no object can be without a subject, so also no subject can be without an object; i.e. no cognizing without something different from it, which is cognized. (Schopenhauer 2005, 733; my emphasis)

It is clear that self-awareness in the sense of being aware of oneself as an autonomous subject is essential for any conscious, mobile animal being. In order to comprehend the world and navigate it successfully, it is necessary for an animal to comprehend their status as an autonomous subject distinct from the outside world. If this form of self-awareness is not present, a being is either (still) absolutely dependent on the care and nurturing of another (think, for example, of a newborn human) or, in the case of a pathological condition, severely impaired and unable to survive without intensive care.

It is generally assumed that there are different degrees of self-awareness; "neither rationality nor self-awareness is an all-or-nothing trait" (DeGrazia 2006, 42) (cf. also Pauen 2007, 145). In order to test whether nonhuman animals possess self-awareness similar to human self-awareness, the mirror test is often used. As discussed earlier in this book, the mirror test can be viewed critically, as it has been developed to test humans, relying on their species-typical abilities, and thus also heavily on sight as a dominant sense. Consequently, the mirror test might not be suitable to test self-awareness in beings belonging to other species, who have other dominant senses: dogs for example, have smell as their dominant sense, rather than sight. Martin Balluch, at a presentation on his book *Der Hund und sein Philosoph* ("*The Dog and his Philosopher*") (2015), for example, remarked that it could be concluded from the fact that a dog recognizes their own smell and can distinguish it from the smells of other dogs that they are self-aware. Similar findings now exist concerning the self-awareness of geckos (Szabo and Ringler 2023).

The mirror test uses specifically human yardsticks for its measurements, which possibly cannot or should not be used as yardsticks for examining and assessing nonhuman animal species. Furthermore, animal species that do not have instruments of touch and manipulation corresponding to human hands cannot logically be expected to use their "hands" to touch the coloured markings. Thus, there are certain biases and methodological limitations of the mirror test.

Passing the mirror test is interpreted as sufficient evidence to infer the existence of self-awareness. However, failing to pass it does not allow the inference that there is no self-awareness. Passing the mirror test is therefore not a necessary condition for the existence of self-awareness (cf. Binder 2014). The nonhuman species that have passed the mirror test largely include species that belong to the great ape group, as well as whales, especially dolphins, and also elephants and magpies.[3] Recent studies have shown that cleaner fishes, too, can identify themselves in a mirror (Kohda et al. 2023). Furthermore, there are research results showing that even

[3] For further elaborations on the mirror test, cf. e.g. Balluch (2005, 112f).

ants marked with a coloured dot and placed in front of a mirror try to wipe the dot off (Cammaerts and Cammaerts 2015, 521).

Pauen (2007, 148) defines self-awareness as follows:

> Self-aware persons possess, firstly, the ability to attribute mental or physical characteristics to themselves. A fundamental prerequisite for this is the recognition of differences [...] between one's own feeling, perceiving, thinking and acting, as well as the feeling, thinking and perceiving of other persons, in short: the ability to put oneself in the perspective of another. Secondly, an idea of one's own characteristics, a self-concept, as well as an autobiographical memory of one's own life story are important. Thirdly, consistency and stability of the self-concept as well as the characteristics of a person are necessary.

It is difficult to verify whether nonhuman animals (can) attribute mental qualities to themselves. As Lori Marino, a neuroscientist, expert in animal behaviour and intelligence, and founder and executive director of The Kimmela Center for Animal Advocacy, notes: "Other animals may reflect on themselves to a much greater degree than we humans do—the simple truth is: we just don't know" (Brensing 2013, 62).

"Are nonhuman animals self-aware? There is reliable evidence that at least some are" (Singer 1994, 148). Singer (ibid.) describes the research findings of Allen and Beatrice Gardner, two US scientists who raised a baby chimpanzee as if she was a human baby without vocal cords: "They communicated with her and with each other in her presence using American Sign Language." The chimpanzee, called "Washoe," lived from 1965 to 2007. She understood about 350 different signs and used about 150 of them in sentences she formulated herself. "As far as self-awareness is concerned, when Washoe is shown her own reflection and asked, 'Who is this?' she does not hesitate to reply, 'I, Washoe'" (ibid.). Washoe thus knew what she looked like and understood that the mirror reflected her own image.

Awareness of Oneself as a Being with a Past and a Future

Peter Singer (1994, 149) states that "[a]pes use [...] signs to refer to past or future events; this shows that they have a sense of time." He (ibid., 149f) recounts an event at the Chimpanzee and Human Communication Institute, run by researchers Roger and Deborah Fouts that supports this statement:

> The sense of time is even more impressive in the festivities that the Fouts couple regularly organise in Ellensburg. Every year after Thanksgiving, they put up a Christmas tree and hang edibles on it. The chimpanzees use the sign combination "candy tree" for the Christmas tree. In 1989, snow fell immediately after Thanksgiving, but the tree was not yet up; so Tatu the chimp asked, "Candy tree?"

In summary, it can be stated that nonhuman animals have already been observed acting in a practically rational manner. Whether they also act theoretically rationally cannot (yet) be determined; possibly research is not yet far enough advanced, and perhaps one day a machine will be invented that makes it possible to read (the) thoughts (of animals), making it possible for one to establish whether nonhuman animals are also theoretically rational. Can nonhuman animals think? Yes, definitely. Some also deceive, which requires complex thought processes. Washoe recognizes herself in the mirror: this is an ability that indicates a human-like type of self-awareness. Finally, it must be asked whether nonhuman animals (can) conceive of themselves as the same thinking beings at different times and places. It seems that we cannot deny that nonhuman animals are also spatio-temporal beings, and that they are also aware of this to a certain extent. It can be assumed that nonhuman animals remember the past and anticipate future events. Also insects take flight in life-threatening situations or try to defend themselves in order to prevent what they fear may be imminent harm or death (Barron and Klein 2016a, b). They can therefore (at least to some extent) anticipate possible (imminent) future events and thus try to prevent them.

We can assume that there are nonhuman animals who exhibit all four of Locke's required characteristics, at least to some extent. In the case of chimpanzees such as Washoe, it is obvious and relatively easy for humans to see that they have a personal identity and should correctly be considered persons. Washoe, by the way, is not only a clever student but also a clever teacher: "Washoe later came to Ellensburg, Washington, where she lived with other chimpanzees under the care of Roger and Deborah Fouts. Here she adopted a baby chimpanzee and soon began not only to use signs towards it, but even to consciously teach it signs" (Singer 1994, 148f).

Locke's definition of "person" does not automatically exclude nonhuman animals. The fact that nonhuman animals can also be persons (as already indicated) possibly entails certain implications that are ethically and legally relevant. In order to avoid normative implications that some humans may find uncomfortable (e.g. banning the killing of nonhuman animals in order to consume and use their bodily products and bodies), one could try to simply increase the requirements for "persons" in order to exclude nonhuman animals once again:

> An astonishingly large number of theorists cling [...] to the hope that personhood can be invoked to reject inviolable rights for (all) animals while maintaining the claim to inviolable rights for (all) humans. In order to hold on to this illusory hope, these theorists engage in increasingly vexatious mental contortions to defend the privileging of humans. (Donaldson and Kymlicka 2013, 71)

A side effect that is relatively unpleasant for speciesist efforts of this kind (for the strength and persuasiveness of the argumentation) is that such an increase in requirements also creates a larger group of human individuals who no longer fulfil the requirements that are (definitely) placed on persons.

The fact that the personhood of certain nonhuman animals is not yet legally recognised goes hand in hand with a means of dealing with them that produces suffering and injustice. In the following, some projects are briefly described that strive for a recognition of the personhood of members of diverse nonhuman species or diverse nonhuman animals. First, is the "The Great Ape Project," initiated by Paola Cavalieri and Peter Singer in 1993, which advocates for granting great apes—such as chimpanzees, gorillas, and orangutans—basic rights akin to those of humans. This includes the right to life, liberty, and protection from torture. The project aims to expand the moral community to recognize these beings as persons deserving of corresponding legal protection. Despite facing challenges and criticism for focusing solely on great apes, the project has successfully influenced legislative changes in Spain and New Zealand. Subsequently, I will then examine the declaration of whale rights, published by the Helsinki Group, and The Nonhuman Rights Project by Steven Wise, before I exemplify the right to a human guardian, legal remedy, and an indirect right of legal action narrating the case of Matthias Hiasl Pan.

12.1.2 The Great Ape Project by Cavalieri and Singer

In an anthropocentric society, because of their similarity to humans, great apes have the best chance of having their status as persons recognised and of being granted basic rights (such as the right to life, freedom and protection from torture). With their Great Ape Project, Paola Cavalieri and Peter Singer proposed an animal rights declaration for great apes (including chimpanzees, orangutans, and gorillas) as early as 1993 (Cavalieri and Singer 1993). They (ibid., 12f) formulated this declaration as follows:

> Declaration on the Great Apes
>
> We demand that the community of equals be expanded to include all great apes: Humans, Chimpanzees, Gorillas and Orangutans.
>
> "The community of equals" is the moral community within which we recognise certain moral principles or rights that govern our relations with each other and which are enforceable in a court of law. These principles or rights include the following:
>
> 1. The right to life
>   The lives of members of the community of equals shall be protected. Members of the community of equals may not be killed except in strictly defined situations such as self-defence.
> 2. The protection of individual freedom
>   Members of the community of equals may not be arbitrarily deprived of their liberty; if they should be imprisoned without prior due process of law, they have the right to immediate release. The imprisonment of those who have not been convicted of any crime or who have not reached the age of criminal responsibility is permitted only if it can be shown that it is for their own good or becomes necessary to protect the community from a member of the community who, if free, would clearly pose a danger to others. In such cases, members of the community of equals have the right to apply to a court either directly or, if they lack the necessary skills, through legal counsel.
> 3. The prohibition of torture
>   To knowingly inflict serious pain on a member of the community of equals, either maliciously or for the alleged benefit of others, is considered torture and is wrong.

Granting Great Apes these three basic rights to life, liberty, and protection from torture would have the consequence that experiments on apes and keeping apes in zoos, for example, would have to be banned, unless those experiments or the keeping of them in zoos were in their interests and for the benefit of their wellbeing (Benz-Schwarzburg 2012, 235). Singer sensibly acknowledges (ibid., 236) that those individuals who could not survive in the wild should be allowed to continue living in protected habitats.

However, if habitats in the wild should no longer be possible, then reserves also represent suitable conditions in which it would be possible for great apes to live well and happily in sufficient freedom (Benz-Schwarzburg 2012, 236).

In 2011, Cavalieri and Singer were awarded the Ethics Prize of the Giordano Bruno Foundation for the Great Ape Project. This award revived the project, which had begun in 1993, so that in 2014 a petition was again submitted to the German Bundestag with the contents of the Great Ape Declaration proposed by the Great Ape Project. In 2015, it was, however, rejected (Great Ape Project n.d.). Although the Great Ape Project thus suffered another setback, it can claim some influence, especially on legislative changes in Spain and New Zealand, as successes. In New Zealand, and now the entire EU—thanks to the European Parliament's directive 2010/63/EU (European Union 2010)—experiments on great apes (amongst others) are now only permitted if they are for the benefit of the experimental animals' own species, or if certain other exceptional circumstances exist:

> The use of great apes, as the species most closely related to humans and with the most highly developed social and behavioural skills, should be permitted solely for research purposes that are conducive to the conservation of those species or where action is justified in relation to a life-threatening or debilitating condition that endangers humans, and where no other species or alternative methods would suffice to adequately fulfil the purposes of the procedure. (ibid.)

Despite its eventual partial success, the Great Ape Project has received negative criticism for its focus on great apes. This focus also has pragmatic reasons—as a first step towards nonhuman animal rights; as the first step that seems most possible and feasible (Cavalieri and Singer 1993, 5).

> So, if we think humans should be granted rights on the basis of possessing these characteristics, other animals who possess them should be granted rights too. Once this is established for great apes, the argument could and should be extended to other species on similar grounds. It would be arbitrary to extend rights just to nonhuman primates, [...] the negative rights that are proposed matter to many species. (Meijer 2013, 44)

Brunn (2015) criticizes this focus and accuses the Great Ape Project of not wanting to break down the human-nonhuman-animal dualism and of shoring up an unjust system: "This stabilizing and reproducing of a dualism of 'equal' and 'unequal' [...] is not progress." Although Brunn (ibid.) would like to see a move away from speciesism as soon as possible, she nevertheless ultimately concedes that ethical norms have also been established *gradually* in the past (ibid.).

12.1.3 Whale Rights

In 2010, the so-called Helsinki Group, formed around Paola Cavalieri, proposed a Declaration on the Rights of Whales. Similar to the Declaration on the Rights of Great Apes, it includes the three basic rights to life, liberty, and protection from torture. It also states that cetaceans cannot and must not be owned by anyone (point 4) and that their habitat and cultural-social structures must not be disturbed (points 5 and 6):

> Declaration of Rights for Cetaceans: Whales and Dolphins Based on the principle of the equal treatment of all persons;
>
> Recognizing that scientific research gives us deeper insights into the complexities of cetacean minds, societies and cultures;
>
> Noting that the progressive development of international law manifests an entitlement to life by cetaceans;
>
> We affirm that all cetaceans as persons have the right to life, liberty and wellbeing. We conclude that:
>
> 1. Every individual cetacean has the right to life.
> 2. No cetacean should be held in captivity or servitude; be subject to cruel treatment; or be removed from their natural environment.
> 3. All cetaceans have the right to freedom of movement and residence within their natural environment.
> 4. No cetacean is the property of any State, corporation, human group or individual.
> 5. Cetaceans have the right to the protection of their natural environment.
> 6. Cetaceans have the right not to be subject to the disruption of their cultures.
> 7. The rights, freedoms and norms set forth in this Declaration should be protected under international and domestic law.
> 8. Cetaceans are entitled to an international order in which these rights, freedoms and norms can be fully realized.
> 9. No State, corporation, human group or individual should engage in any activity that undermines these rights, freedoms and norms.
> 10. Nothing in this Declaration shall prevent a State from enacting stricter provisions for the protection of cetacean rights.
>
> Agreed, 22nd May 2010, Helsinki, Finland. (Cetacean Rights 2010)

The fact that whales and their habitat (point 5 in the above declaration) should be better protected is urgently necessary, considering the massive negative influence that humans exert on them. The oceans are heavily impacted and damaged by shipping, interventions for economic purposes, pollution, contamination and overfishing: "[T]he fishing grounds of the North Atlantic and the Mediterranean are now practically empty due to the gigantic trawl nets" (Greenpeace n.d.). In particular, economic interventions by companies exploring marine mineral resources with the help of so-called airguns, and the loud noise produced by installation of wind farms in the sea, as well as shipping traffic, cause massive noise pollution in the world's oceans, as noted by Karsten Brensing (walrecht de 2013a, b). This noise pollution impacts whales considerably, since in addition to the acoustic torment this noise causes them, it severely disrupts their communication and orientation abilities (ibid.). This noise pollution causes whales to strand more often, and renders them unable to communicate with possible mates; furthermore, many of them die as by-catch through the use of trawl nets for fishing (whose use is actually prohibited) (ibid.). In consequence, there are fewer individuals, and fewer mating pairs, which means a further endangerment of the population (ibid.).

The demand for a right not to be kept in captivity (point 2 of the above declaration) is based on the fact that, as already mentioned, keeping cetaceans in captivity in an ethically justifiable manner is impossible, due to the large areas they require for their natural behaviour and complex social structures.

12.1.4 The Nonhuman Rights Project

The Nonhuman Rights Project, founded in 2007 by US legal scholar Steven M. Wise, attempts to advance the animal rights movement by legal means. The historical case of the US-American slave James Somerset, who escaped from enforced servitude in the US but ended up in court in England as an "escaped slave," served as inspiration for Wise, the project he initiated, and the associated organization he built. In the case of James Somerset in eighteenth-century England, the judge, Lord Mansfield, decided that Somerset must be justly set free. This is seen as a precedent for the abolition of slavery in the US (Wise 2005). The aim of the Nonhuman Rights Project is to achieve a "Lord Mansfield ruling" on nonhuman animal rights as a precedent for further lawsuits. The Nonhuman Rights Project seeks to win rights and recognition of the personhood status of certain nonhuman animals. In particular, the project represents members of those species that are scientifically (recognised as) self-aware and autonomous beings. These are, for example, great apes, elephants, and whales. By achieving the recognition of the personhood of nonhuman individuals, the Nonhuman Rights Project seeks to secure rights such as bodily liberty and bodily integrity for them.

Tommy, a chimpanzee, was the first "client" of the Nonhuman Rights Project. A short summary of his story is as follows:

> When we found Tommy, a male chimpanzee, he was living alone in a cage in a shed on a used trailer lot along Route 30 in Gloversville, New York. Believed to have been born in the early 1980s, Tommy was raised from infancy by Dave Sabo, former proprietor of "Sabo's Chimps". Tommy appeared as "Goliath" in the 1987 film Project X, according to Sabo. Animal activist and TV icon Bob Barker and others alleged that trainers beat the chimpanzees used in the film with blackjacks and clubs. After Sabo died in 2008, "ownership" of at least some of the chimpanzees passed to the Laverys. The status of these other chimpanzees—including whether or not they're still alive—is unknown. Concrete walls painted to look like a jungle were the only intimation of a chimpanzee's natural habitat in Tommy's cage. A television was his only company. "He'd rather be by himself," Patrick Lavery told the Albany Times Union in 2013. "He likes being by himself." Before our litigation concluded, media outlets began to report that Tommy had disappeared from Gloversville. His owner disingenuously claimed in interviews not to remember where he sent him, but private investigations conducted by the NhRP and federal and state records requests showed that Tommy had been moved out of state in 2015. In 2022, we announced our belief that Tommy was imprisoned in a roadside zoo in Michigan called the DeYoung Family Zoo. As part of our efforts to secure his release to a sanctuary, over 3000 NhRP supporters called on the Michigan Department of Agriculture & Rural Development (MDARD) to conduct an emergency inspection at the zoo. Behind the scenes, we were preparing new litigation on behalf of Tommy, whose right to liberty we had never given up on. In December of 2023, in the midst of our final preparations for this case, we were devastated and outraged to learn Tommy had died. USDA records sent to us on Nov. 30, 2024 included an exchange between representatives from MDARD and the USDA in response to the emails NhRP supporters sent with their concerns about Tommy. The exchange suggests that Tommy died

> much as he had lived: alone, vulnerable, with no comfort but what he could offer himself. He had apparently been found dead in February of 2022, "curled up in his sleeping spot" inside a building on the DeYoung Family Zoo property, according to a USDA official. Tommy's death shows us how animal welfare fails nonhuman animals—everything that happened to him was legal. (Nonhuman Rights Project 2024b)

The Nonhuman Rights Project has received support from academics in the form of briefs written by amici curiae philosophers in several court cases, including an amici curiae philosophers' brief on behalf of Tommy (Andrews et al. 2018). The Nonhuman Rights Project also challenged the lawfulness of the captivity of the elephant Happy, seeking habeas corpus relief and a transfer to a sanctuary: "Torn from her natural habitat and herd in Thailand when she was a baby, Happy has been held captive in the Bronx Zoo for almost half a century" (Nonhuman Rights Project 2024a). Another amici curiae philosophers' brief (Comstock et al. 2020) detailed the ways in which captivity was not compatible with Happy's personhood; unfortunately, as in the case of Tommy, Happy's story is also missing a happy ending:

> From 2006 to the present, to protect Happy from the other elephants and with assurances from zoo director Jim Breheny that Happy is sufficiently happy where she is, "The Bronx Zoo's Loneliest Elephant" has lived alone, without a true elephant companion, in a rotating portion of the 1.15-acre exhibit. According to *The New York Post*, "Happy spends most of her time indoors in a large holding facility lined with elephant cages, which are about twice the length of the animals' bodies. The public never sees this. (Nonhuman Rights Project 2024a)

Thanks to the important work of animal rights organizations like the Nonhuman Rights Project, we may hope that the situation nonhuman animals find themselves in this world will gradually improve. In the following chapter, I provide reflections on a nonhuman animal right to a human guardian, a legal remedy, and an indirect right of legal action. I first examine efforts to legally establish guardianship and personhood for the chimpanzee Mathias Hiasl Pan in Austria. Activists sought guardianship to prevent Hiasl's sale to circuses or laboratories after the shelter he lived in went bankrupt. They argued that Hiasl deserved personhood due to biological and cognitive similarities to humans, citing Kantian criteria; despite legal hurdles, the case drew global attention. I then also discuss animal rights advocate Gary Francione's stance against the legal status of nonhuman animals as property. He criticizes existing animal welfare laws, arguing that they fail to adequately protect nonhuman animal interests. I then provide an empirical discussion and show that in Austria, too, despite some noteworthy efforts—which include animal welfare ombudspersons in every province—the focus remains on reducing animal suffering rather than on taking individual animal interests seriously.

References

Andrews, Kristin, Gary Comstock, Gillian Crozier, Sue Donaldson, Andrew Fenton, Tyler John, L. Syd M. Johnson, et al. "The Philosopher's Brief on Chimpanzee Personhood." Proposed brief by amici curiae philosophers in support of the petitioner-appellant, Court of Appeals, State of New York, 2018.

Aristotle. *Aristotle in 23 Volumes: Volume 21: Politics*. Translated by Harris Rackham. Cambridge, MA: Harvard University Press, 1944.

Balluch, Martin. *Die Kontinuität von Bewusstsein: Das naturwissenschaftliche Argument für Tierrechte*. Vienna: Guthmann-Peterson, 2005.

Balluch, Martin. Presentation given about book *Der Hund und sein Philosoph: Plädoyer für Autonomie und Tierrechte*. University of Salzburg. November 5, 2015.

Barron, Andrew B., and Colin Klein. "What insects can tell us about the origins of consciousness." *Proceedings of the National Academy of Sciences of the United States of America* 113, no. 18 (2016a): 4900–4908. DOI:https://doi.org/10.1073/pnas.1520084113.

Barron, Andrew B., and Colin Klein. "Insects have the capacity for subjective experience." *Animal Sentience* 1, no. 9 (2016b): 1–19. DOI: https://doi.org/10.51291/2377-7478.1113.

Benz-Schwarzburg, Judith. *Verwandte im Geist – Fremde im Recht – Sozio-kognitive Fähigkeiten bei Tieren und ihre Relevanz für Tierethik und Tierschutz*. Erlangen: Harald Fischer, 2012.

Binder, Alexandra. "Auch wer sich im Spiegel nicht erkennt, kann Ich-bewusst sein." Vet-Journal (February 2014): 46–49. Accessed October 11, 2015. veterinaerjournal.at/fileadmin/userupload/PDF/V0214/V02_REPORTAGE.pdf

Blattner, Charlotte E., Kendra Coulter, and Will Kymlicka. "Animal Labour and the Quest for Interspecies Justice." In *Animal Labour. A New Frontier of*

Interspecies Justice? edited by Charlotte E. Blattner, Kendra Coulter, and Will Kymlicka, 1–25. Oxford: Oxford University Press, 2020.

Brensing, Karsten. *Persönlichkeitsrechte für Tiere.* Freiburg im Breisgau: Herder, 2013.

Brunn, Franziska. "Wer wie wir ist, bekommt Rechte?!" Accessed November 1, 2015. tierbefreiung.de/archiv/56/gap_brunn.html.

Cammaerts, Marie-Claire, and Roger Cammaerts. "Are Ants (Hymenoptera, Formicidae) Capable of Self Recognition?" *Journal of Science* 5, no. 7 (2015): 521–532.

Castelló, Pablo P. "The Language of Zoodemocracy: Contesting Human Sovereignty Over Animals." PhD thesis. Royal Holloway, University of London, 2022a.

Castelló, Pablo P. "The erasures of Peter Singer's theory, and the ethical need to consider animals as irreducible others." *Philosophy Today*, 66, no. 1 (2022b): 637–653. DOI: https://doi.org/10.5840/philtoday2022325454.

Cavalieri, Paola, and Peter Singer, eds. *Menschenrechte für die Großen Menschenaffen: "Das Great Ape Project."* Munich: Wilhelm Goldmann Verlag, 1993.

Cetacean Rights. "Declaration of Rights for Cetaceans: Whales and Dolphins." Published May 22, 2010. Accessed November 18, 2015. https://www.cetaceanrights.org.

Comstock, Gary, G. K. D. Crozier, Andrew Fenton, Tyler John, L. Syd M. Johnson, Robert C. Jones, Nathan Nobis, David M. Peña-Guzmán, et al. "The Philosophers' Brief on Elephant Personhood." New York State Appellate Court, 2020.

Conradi, Elisabeth, and Eva Stollreiter. *Brockhaus Philosophie: Ideen, Denker und Begriffe*. Mannheim: F.A. Brockhaus, 2004.

Deckha, Maneesha. *Animals as Legal Beings: Contesting Anthropocentric Legal Orders.* Toronto: Toronto University Press, 2020.

Deckha, Maneesha. *Animals as Legal Beings: Contesting Anthropocentric Legal Orders.* Toronto: University of Toronto Press, 2021.

DeGrazia, David. (2006) "On the Question of Personhood beyond Homo Sapiens." In *In Defense of Animals: The Second Wave*, edited by Peter Singer, 40–53. Oxford: Blackwell Publishing, 2006.

Donaldson, Sue, and Will Kymlicka. *Zoopolis: Eine politische Theorie der Tierrechte.* Translated by Joachim Schulte. Berlin: Suhrkamp, 2013.

Eisen, Jessica. "Down on the Farm: Status, Exploitation, and Agricultural Exceptionalism." In *Animal Labour: A New Frontier of Interspecies Justice?* edited by Charlotte Blattner, Kendra Coulter, and Will Kymlicka, 139–159. Oxford: Oxford University Press, 2020.

European Union (EU). "Richtlinie 2010/63/EU des Europäischen Parlaments und des Rates vom 22. September 2010 zum Schutz der für wissenschaftliche Zwecke verwendeten Tiere." Published October 20, 2010. Accessed October 11, 2015. eur-lex.europa.eu/LexUriServ/LexUriServ.do?uri=OJ:L:2010:276:0033:0079:de:PDF.

Great Ape Project (GAP). "GAP Relaunch." Published n.d. Accessed October 11, 2015. greatapeproject.de/relaunch.

Greenpeace. "Fischerei-Industrie: Leere Meere – volle Kassen." Published n.d. Accessed November 19, 2015. greenpeace.de/themen/meere/fischerei.

Kohda, Masanori, Redouan Bshary, Naoki Kubu, Satoshi Awata, Will Sowersby, Kento Kawasaka, Taiga Kobayashi, and Shumpei Sogawa. "Cleaner fish recognize self in a mirror via self-face recognition like humans." *Proceedings of the National Academy of Sciences of the United States of America*, 120, no. 7 (February 2023) e2208420120. DOI: https://doi.org/10.1073/pnas.2208420120.

Locke, John. *Versuch über den menschlichen Verstand.* Band I: Buch I und II. Translated by Carl Winckler. Hamburg: Meiner, 2000.

Meijer, Eva. "Political Communication with Animals." *Humanimalia* 5, no. 1 (2013): 28–52.

Nonhuman Rights Project (NhRP). "Happy." Published n.d.-a. Accessed April 19, 2024a. https://www.nonhumanrights.org/client/happy/.

Nonhuman Rights Project (NhRP). "Tommy." Published n.d.-b. Accessed April 19, 2024b. https://www.nonhumanrights.org/client/tommy/.

Pauen, Michael. *Was ist der Mensch? Die Entdeckung der Natur des Geistes.* Munich: DVA, 2007.

Pauketat, Janet and Jacy Anthis. "Predicting the moral consideration of artificial intelligences." *Computers in Human Behavior.* 136 (2022): 107372. DOI: https://doi.org/10.1016/j.chb.2022.107372.

Precht, Richard David. *Noahs Erbe – Vom Recht der Tiere und den Grenzen des Menschen.* Reinbek bei Hamburg: Rowohlt Taschenbuch Verlag, 2000.

Schopenhauer, Arthur. *Die Welt als Wille und Vorstellung.* Paderborn: Voltmedia, 2005.

Shoemaker, Sydney. "Persons, Animals, and Identity." Synthesis 162, no. 3 (2008): 313–324. DOI: https://doi.org/10.1007/s11229-007-9253-y.

Singer, Peter. *Praktische Ethik.* Translated by Oscar Bischoff, Jean-Claude Wolf, Dietrich Klose, and Susanne Lenz. Stuttgart: Reclam, 1994.

Singer, Peter. *The Expanding Circle.* Princeton, NJ: Princeton University Press, 2011.

Szabo, Birgit, and Eva Ringler. "Geckos differentiate self from other using both skin and faecal chemicals: evidence towards self-recognition?" *Animal Cognition* 26, no. 3 (2023): 1011–1019. DOI: https://doi.org/10.1007/s10071-023-01751-8.

Tanasescu, Mihnea. (2014) "The Rights of Nature: Theory and Practice." In *Political Animals and Animal Politics*, edited by Marcel Wissenburg and David Schlosberg, 150–163. London: Palgrave Macmillan, 2014.

Wallace, R. Jay. "Practical Reason." In *The Stanford Encyclopedia of Philosophy Archive: Summer 2009 Edition*, edited by Edward N. Zalta and Uri Nodelman. Last modified 2009. http://plato.stanford.edu/archives/sum2009/entries/practical-reason/.

walrecht de. "Karsten Brensing gemeinsam mit Veronica Ferres umd Alicia von Ritterberg bei Markus Lanz." Uploaded to YouTube April 23, 2013a. Accessed November 19, 2015a. youtube.com/watch?v=caiMyMZBoj0.

walrecht de. "Karsten Brensing gemeinsam mit Veronica Ferres umd Alicia von Ritterberg bei Markus Lanz 2/2." Uploaded to YouTube April 23, 2013b. Accessed November 19, 2015b. https://www.youtube.com/watch?v=_xogvPhDWYs.

Wise, Steven M. *Though the Heavens May Fall: The Landmark Trial That Led to the End of Human Slavery*. Cambridge, MA: Perseus Books, 2005.

CHAPTER 13

Specific Rights in Nonhuman Animal Rights Declarations (Part 3)

13.1 Article 6: Right to a Human Guardian, a Legal Remedy, and an Indirect Right of Legal Action

13.1.1 *The Case of Mathias Hiasl Pan*

On 6 February 2007 an application for guardianship of a man named Mathias Hiasl Pan, born in 1981 in Sierra Leone, and living in Vösendorf was filed at the District Court of Mödling, Lower Austria (Balluch 2014, 84ff). When the judge read Mr. Pan's residential address, she became suspicious, because the residential address was the address of an animal shelter in Vösendorf (ibid.). Mr. Pan was, in fact, not a *Homo sapiens*, but a *Pan troglodytes*, i.e. a chimpanzee. However, the initiators of the application for guardianship—Martin Balluch and his colleagues—were quite serious in their application. They wanted to have the chimpanzee Hiasl, who had initially been bought by a pharmaceutical company as a laboratory animal, but was, as of 2006, housed in the animal protection centre in Vösendorf, which had had to file for bankruptcy, placed under guardianship (ibid.). Hiasl's welfare was directly threatened at this time: he was to be sold as part of the bankrupt estate, potentially either to a foreign circus—since a ban on wild animals in circuses in Austria has been in force

D. Schneeberger, *Envisioning a Better Future for Nonhuman Animals*, The Palgrave Macmillan Animal Ethics Series, https://doi.org/10.1007/978-3-031-75332-9_13

since 2005 (Rechtsinformationssystem des Bundes n.d.-b, § 27)—or to an animal experimentation laboratory abroad—as, again, in Austria, experiments on great apes, for purposes other than their inherent interests or their own wellbeing, have been prohibited since 2006 (ibid. n.d.-c, § 13). He therefore stood not only to be sold, but possibly even killed. A group of animal rights activists tried to source the living costs for Hiasl; these, however, could only be donated to the animal shelter in Vösendorf, which meant that they would then directly go to the creditors, rather than reaching Hiasl; in order to be able to help Hiasl, Balluch and his colleagues thus applied for guardianship of Hiasl (Balluch 2014, 84ff).

A first oral hearing followed, which ended with the judge stating that Balluch and his colleagues had good but misguided intentions (ibid., 84). She rejected the application for guardianship on the grounds that no documents had been submitted by Mr. Pan. However, under the argumentation advanced by Balluch and colleagues, this reasoning did not pose a problem: there had already been human precedents where (due, for example, to the chaos of war) no further proof of identity could be presented. Comparable circumstances applied to Hiasl, since he had been abducted in Sierra Leone at the age of ten months and subsequently brought to Vienna (ibid., 71f).

"In order for Hiasl to be granted guardianship, he had to be considered a person under civil law. In Austria, § 16 of the General Civil Code defines the concept of a person indirectly. According to the prevailing legal opinion, only human beings could be natural persons" (ibid., 84). The task of the animal rights activists was now to argue that Hiasl was a person. Three expert opinions were attached to support this claim. The first expert opinion was provided by Professor Volker Sommer, an internationally recognised expert in evolutionary anthropology. He argued that Hiasl, as a chimpanzee, could be categorized biologically as a human being: "This on the one hand because the species concept has no biological reality and on the other hand because the human concept is biologically a name for a genus (Homo), and according to common nomenclature it could well be scientifically argued that the chimpanzee as *Homo pans* [sic!] belongs to it" (ibid.). Sommer also illustrated (ibid., 85) the biological proximity between humans and chimpanzees by noting that hybrids between humans and apes would likely be biologically possible "if human males mated with captive ape females, especially chimpanzee females," since the biological proximity between humans and the (remaining) great apes is at least as close as that between horses and donkeys, for example. Sommer also argued that there is no biological "species-specific feature" that distinguishes humans from the great apes (ibid.).

The second expert opinion was offered by Signe Preuschoft, a high-profile expert on chimpanzees and the scientific director of the chimpanzee rehabilitation programme in Vösendorf: "The second argument we made was about the concept of persons in civil law. We argued that persons, in the Kantian sense, are those beings who recognize the personhood of others" (ibid., 86). Preuschoft argued that chimpanzees would fulfil Kant's criterion to be a person (ibid.): they can be recognised as "intentional systems"; thus, intentions can be attributed to them and, moreover, they can also recognize intentions in other beings. Preuschoft noted: "Chimpanzees, for example, seem to see a difference in not giving them something because you were clumsy or because you were unwilling. In the future they will try to continue to cooperate with the one who has been mishandled, but not with the one who has been uncooperative" (ibid.).

"The third expert opinion was written by two university professors of the Vienna Juridicum [law school], Eva-Maria Maier and Stefan Hammer" (ibid., 87). Their expert opinion was about the fact that animals are legally already "not things," according to § 285a of the Austrian Civil Code (ABGB): "If one understands § 285a ABGB as a legal recognition of an unconditional [...] value of animals, this rather refers to a possible expansion of the circle of natural legal subjects by adding animals as further 'natural legal comrades' to humans as 'natural persons'" (ibid., 87).

With these three expert opinions, the animal rights activists around Balluch attended the second oral hearing at the Mödling District Court. The judge decided not to appoint a guardian for Hiasl, "because he does not need one" (ibid., 88). An appeal was then filed with the Vienna Regional Court. "The head of the court called my lawyer and threatened that he would report him to the bar association for contempt of court and conduct unbecoming of a lawyer if he did not withdraw his applications by 6 p.m. the same day. Apparently they wanted to prevent a legal dispute about Hiasl's personhood status by all means" (ibid., 88f). The application for recourse was not withdrawn. A little later, however, it was rejected by the Vienna Regional Court on the grounds that "this appeal could only have been brought by the person concerned himself or his trustee" (ibid., 89). It is evident that this reasoning is logically and (thus necessarily also) legally problematic. As a final step, Balluch and his colleagues brought the case before the European Court of Human Rights (ECHR). The animal rights activists' justification for bringing the case before the ECHR was that Hiasl had not been given a fair trial and the question of his status as a person had not been investigated. However, the application for the case to be dealt with by the ECHR was rejected in early 2008 on the grounds that it was "none of Mr. Balluch's business whether Hiasl was given a guardian or not" (ibid., 89).

The case of Matthias Hiasl Pan made headlines around the world, and provoked considerable discussion on the topic. In the end, Hiasl was spared any further upset, as the Vienna Animal Protection Centre was rescued from bankruptcy and Hiasl did not have to be sold. Balluch (ibid., 90) emphasizes: "The difference between a thing and a person according to the law is huge. Things are property but can't have property, conversely persons are allowed to have property but can't be property."

Gary Francione in particular is decidedly against the practice (and convention) of nonhuman animals being seen and treated as property (2007, 257). For Francione, this categorization goes hand in hand with the fact that too little attention is paid to the interests of nonhuman animals (if they play any role at all); writing on his Facebook page—"Gary L. Francione: The Abolitionist Approach to Animal Rights"—in 2015, Francione opined: "I maintain that, because animals are property, the law [...] is structurally unable to protect their interests in any significant way." Although nonhuman animals are legally no longer things in Austria, they are nevertheless treated as such, if there are no further special provisions for dealing with them in certain situations: "The regulations applicable to things are to be applied to animals only insofar as no deviating regulations exist" (Rechtsinformationssystem des Bundes n.d.-d; § 285a).

The Great Ape Project, the Convention on the Rights of Whales, the Nonhuman Rights Project, and the case of Mathias Hiasl Pan push for establishing better protections for nonhuman animals. Living beings who are subjects-of-a-life and/or persons, or—given our still considerable ignorance of the consciousness and cognitive capacities of a wide variety of animals—nonhuman animals in general, should not (and here I agree with Francione (2007, 257)) be conceptualized as possessions, even if they live in the care of a human guardian in an interspecies community. The precise designation (personhood, animalhood, beingness) is not the important thing: it is the actual protection that the designation affords.

To exclude nonhuman animals "from the group of things in a legal sense in supplements to the BGB or ABGB, but then to treat them *de facto* legally like things, seems unsatisfactory" (Benz-Schwarzburg 2012, 249). Benz-Schwarzburg (2012, 249f) outlines two possible solutions to this problem. Firstly, one could classify some nonhuman animals as persons on the basis of their species affiliation. Secondly, one could alternatively strive for a fundamental redefinition of the status of certain nonhuman animals (ibid.). Wissenburg (2014, 40) points to a broader spectrum of possibilities for the (legal) categorization of a nonhuman animal: "the either/or dichotomy needs to be replaced by a subtler conception of 'the subject' compatible with a wider range of possible legal statuses for animals." Cochrane (2010, 141) also argues against the conceptualization of nonhuman animals as things or commodities: "If animals are to enjoy their rights, they need to be regarded as sentient individuals who merit justice for their own sake, and not as mere commodities to be raised and killed as efficiently as possible." For nonhuman animals, a third category could be established between things and persons, for example, a subcategory of the concept of person:

> Certainly, great apes, just like human children, are not legal subjects with full legal capacity and therefore it would have to be discussed which concrete, probably limited rights the status of the nonhuman person would entail. However, there is much to suggest that these would also have to be some strong basic rights of protection in the sense of the three rights of life, liberty and absence of torture. After all, we also grant these to humans who are not legally competent. (Benz-Schwarzburg 2012, 250)

In particular, a novel status would have to be created for nonhuman animals that (have to) live as members of interspecies communities. Wissenburg (2014, 42) argues that a status located between property and citizen should be created for nonhuman animals, suggesting taking the concept of property and then modifying it, so that the owner becomes a protector or caretaker: "a curator, a guardian, a warden, a trustee, a shepherd or an advocate" (ibid.).

Next to a nonhuman right to a legal remedy and an indirect nonhuman right of legal action, a nonhuman right to a human guardian is crucial for nonhuman beings who live in multispecies societies with humans and are dependent on a human guardian to flourish in this context. In order to thrive as members of these communities, similar to human children or mentally disabled humans, such nonhuman animals need (at least) one human being who assumes the duty of care for them. This person has various rights and duties towards the individual. Guardians have the task of caring for the individual and making decisions in their best interests for them (cf. Article 3 of the Convention on the Rights of the Child ratified in 1990: "In all actions concerning children [...] the best interests of the child shall be a primary consideration").

Donaldson and Kymlicka (2013, 224) prefer the categorization of nonhuman animals living in interspecies communities with humans as citizens, over a guardianship model. Although Donaldson and Kymlicka (ibid.) think that guardianship and citizenship models lead to similar end results (at least as far as some issues are concerned), they (ibid., 227) prefer the citizenship model, because it conceives of nonhuman animals not as dependent "wards," but as full political actors. For pragmatic reasons—the transition from animal owner to animal guardian is not as far as that from ownership to citizenship—I nevertheless see some advantages in the guardianship model.

Although the duty of care for a nonhuman animal living in an interspecies community with humans initially lies privately with the guardian, the state must intervene in the event that the guardian does not or cannot fulfil their duty of care sufficiently well. In order to be able to protect and guarantee the rights of nonhuman animals living in interspecies communities with humans, the state must assume a certain monitoring and controlling function—cf. again the Universal Convention on the Rights of the Child: "States Parties undertake to ensure to the child such protection and care as is necessary for his or her wellbeing [...], taking into account the rights and duties of his or her parents, legal guardians or other persons legally responsible for the child." Austrian law provides that the person responsible for the care of a nonhuman animal can be deprived of them and that the authority must do so if the "owner" does not sufficiently fulfil their duties of care and does not or cannot correct this even after a warning, according to the Animal Protection Act (Rechtsinformationssystem des Bundes (RIS) n.d.-e; §§ 36 & 37):

> Entering properties, rooms and means of transport, duty to cooperate
>
> § (1) The bodies of the authorities entrusted with the enforcement of this Federal Act and the experts consulted as well as the veterinary experts of the Commission of the European Communities shall have the right to enter properties, premises and means of transport for the purpose of inspection (§ 35), subject to the necessary veterinary precautions, and to gain access to them, subject to the proportionality of the means used, if such

access is not granted voluntarily. This shall also apply if there is a reasonable suspicion that this Federal Act has been violated. The person responsible for keeping the animals shall be given the opportunity to be present during the inspection, provided that the purposes of the inspection are not impaired.

[...]

(3) Persons involved in animal husbandry shall provide the necessary information upon request. [...]

§ 37

Immediate coercion

§ Section 37. (1) The organs of the Authority shall be obliged,

1. to terminate perceived violations of §§ 5 to 7 by direct official command and coercive power;
2. take away from the owner an animal found in a condition which suggests that the animal will suffer pain, suffering, harm or severe distress without immediate remedy if the owner is unwilling or unable to remedy the situation.

Next to a human guardian, nonhuman animals living in multispecies societies with humans who are dependent on human care should also have the right to a name. The bestowal of an individual name is usually accompanied by a degree of esteem and recognition of the individuality of a being, as seen in Article 7 of the Convention on the Rights of the Child: "The child shall be registered immediately after birth and shall have the right to a name from birth." Registration immediately after birth is also a prerequisite for the state to be able to fulfil its duty of control and intervention, if necessary. A registration obligation is already provided for in Austrian law for some domesticated animal species, for example, dogs (Bundesministerium für Soziales, Gesundheit, Pflege und Konsumenschutz n.d.-b), horses and various other animal species (cf. the Austrian Animal Identification and Registration Ordinance (*Bundesgesetzblatt für die Republik Österreich* 2009)).

The person responsible for the care of a nonhuman animal must ensure that the living conditions are appropriate for them. They must also look after the wellbeing of the nonhuman individual and make decisions in their best interest where the nonhuman animal cannot do so alone. If the guardian cannot or does not want to fulfil their task for whatever reason, a new guardian must be found. If no guardian can be found (for the time being), the state must step in. This requires special care institutions (sanctuaries) that can take in nonhuman animals, since abandonment or even killing would no longer be legally possible once a nonhuman animal right to life (for certain special nonhuman animals) were legally established. Although such an idea may still be intuitively unfamiliar to us, there are certain nonhuman animal species for which our intuitions already cater in this fashion.

Ideally, in the future, no "surplus" animals would be killed in animal shelters. The living conditions must also meet the needs of the individuals kept in these shelters and must not run too far counter to those needs. The problem of "surplus" domesticated nonhuman animals who cannot find a place in interspecies communities must of course be tackled at its root through a responsible reproductive policy. Those in charge of the welfare of individuals must thus also be mindful of responsible reproduction choices. Whilst the sexual and reproductive capacities of these nonhuman animals should be valued, respected and, where appropriate, not be hindered, offspring should be prevented if circumstances do not suggest that the offspring can be well cared for and that a good and happy life (in the community) can be provided for them.

In all actions involving a nonhuman animal individual, the welfare of the nonhuman animal must be considered as paramount as possible. If the nonhuman animal has a close bond to a caregiver or to conspecifics in a community and/or the individual is already accustomed to and comfortable in a given environment, relocation may potentially cause (considerable) stress and have a (strong) negative impact on an animal's wellbeing. Relational harms (cf. Cooke 2021; as well as the next chapter in this book) must be avoided where possible. Therefore, translocations from otherwise good living conditions should only be implemented for valid reasons (cf. Austrian Animal Husbandry Ordinance § 2.2: "Any change in the husbandry conditions of an animal in captivity shall be avoided if there is a risk of causing pain, suffering or harm to the animal or putting it in severe fear." Rechtsinformationssystem des Bundes (RIS) n.d.-a). The right to a human guardian and legal remedy could enable nonhuman beings to flourish in a multispecies society. Therefore, the right to a human guardian, a legal remedy, and indirect right of legal action are important future nonhuman animal rights.

Nonhuman animals are currently "protected" to a certain extent by animal protection laws in Western countries, or rather, their use is regulated. However, if one assumes that nonhuman animals actually have a *right to demand that* their "users" comply with the requirements of the

Animal Welfare Act, then logically this right should also be accompanied by a right of indirect legal action.

Since nonhuman animals cannot exercise this right themselves in the human system, they would have to be provided with a legal remedy (similar to children, and mentally impaired or mentally ill humans, or those with dementia). This person would have to be able to file lawsuits on their behalf: "Inasmuch as a right (whether moral or legal) is fundamentally a claim, it is indeed the case that its granting secures, in principle, a kind of representation" (Tanasescu 2014, 153). In Austria, a step in this direction has already been taken with the establishment of the animal protection ombudspersons; in Germany, too, there is a large-scale, if not yet nationwide, right of associations to sue in animal protection matters. Various German animal protection organizations are calling for a nationwide establishment of this right of action (for animal protection associations) (cf. Albert Schweitzer Stiftung n.d.).

Concerning working towards a nonhuman right to a legal remedy and an indirect nonhuman right of legal action, the institution of animal welfare ombudspersons might be an interesting institution to examine. At present, there are animal welfare ombudspersons in every province of Austria. Even though they cannot necessarily provide guaranteed legal remedy, they are an official contact for reporting animal welfare offences. Animal protection ombudspersons must have "a degree in veterinary medicine, zoology or agricultural sciences [...] or comparable training" as well as having completed "additional training in the field of animal protection" (Bundesministerium für Soziales, Gesundheit, Pflege und Konsumenschutz n.d.-a). It should be kept in mind that "animal welfare ombudsmen or ombudswomen have the task of representing the interests of animal welfare" (ibid.; my translation), and not the interests of the nonhuman animals concerned. This is a significant difference.

The animal protection ombudspersons ensure compliance with the provisions of the Animal Protection Act and have party status in "administrative proceedings, including administrative penal proceedings [...] and are entitled to inspect all procedural files and to demand all relevant information. The authorities shall support the animal protection ombudspersons in the exercise of their office" (ibid.; my translation). The ombudspersons are responsible for a diverse range of activities such as:

> Information on regulations, transitional periods, husbandry conditions and minimum requirements under the Animal Welfare Act.

> Counselling on animal husbandry appropriate to the species, the correct procedure in the event of suspected offences against the Animal Welfare Act and referral to the right contact persons.
>
> Public relations work, campaigns, participation in various events and cooperation with animal protection organizations are measures to raise awareness of animal protection and species-appropriate animal husbandry among the population. (ibid.)

Although the institution of these animal protection ombudspersons is inevitably part of the system of using, killing and exploiting nonhuman animals, and those working there have functions that maintain and stabilize the system, it can nevertheless be judged as a step in the right direction that the institution of such specific ombudspersons exists in Austria. Even though it is difficult to assess such an institution from the outside without more detailed knowledge about its actual activities and their effectiveness, it can be hoped that this institution might at least have some positive effects on the lives of some of the nonhuman animals concerned.

In the next chapter, I reflect on the significance of love, sexuality, procreation, family, and friendships for nonhuman animals. I underscore the essential nature of these needs for both humans and nonhuman animals and discuss what Steve Cooke calls "relational harms." I provide biological and ethological insights and examine romantic love as an evolutionary strategy for reproductive success, rooted in mammalian bonding behaviours. I furthermore discuss the widespread occurrence of maternal love and care in the animal kingdom. I explore pair-bonding behaviours observed in geese, dolphins, and voles. Nonhuman animals, such as ducks and geese for example, exhibit lifelong partnerships, reflecting a deep emotional connection and commitment. These behaviours challenge the notion that human experiences of love and sexuality are uniquely human: they are instead evolutionary commonalities shared by many animals. However, human interventions often interfere with nonhuman animals' sexuality, reproductive capabilities, and relationships without duly considering their intrinsic value. I make the case that we need to critically examine our treatment of nonhuman animals regarding their sexual, reproductive rights, and rights to family, and friends. Learning from the discussion concerning the sexual and reproductive rights of disabled humans might inform our thinking when it comes to these rights in the nonhuman animal domain, which is why I also explore these phenomena when it comes to disabled humans in the next chapter.

References

Albert Schweitzer Stiftung. "Verbandsklagerecht im Tierschutz." Published n.d.. Accessed November 1, 2015. https://albert-schweitzer-stiftung.de/themen/verbandsklagerecht.

Balluch, Martin. *Der Hund und sein Philosoph: Plädoyer für Autonomie und Tierrechte*. Vienna: Promedia, 2014.

Benz-Schwarzburg, Judith. *Verwandte im Geist – Fremde im Recht – Sozio-kognitive Fähigkeiten bei Tieren und ihre Relevanz für Tierethik und Tierschutz*. Erlangen: Harald Fischer, 2012.

Bundesgesetzblatt für die Republik Österreich. "291. Verordnung des Bundesministers für Gesundheit über die Kennzeichnung von Schweinen, Schafen, Ziegen, und Equiden." Published September 10, 2009. Accessed November 26, 2015. https://animaldata.com/TKVO.pdf.

Bundesministerium für Soziales, Gesundheit, Pflege und Konsumenschutz. "Schwerpunkt: Tiergesundheit und Tierschutz." Published n.d.-a. Accessed October 11, 2015. bmg.gv.at/home/Schwerpunkte/Tiergesundheit/Tierschutz/Ombudsstellen.

Bundesministerium für Soziales, Gesundheit, Pflege und Konsumenschutz. "Häufige Fragen: Chip-Pflicht, Kennzeichnung und Registrierung von Hunde." Published n.d.-b. Accessed November 26, 2015. https://www.sozialministerium.at/home/Service/FAQ_Haeufige_Fragen_/Chip_Pflicht_Kennzeichnung_und_RRegistrierun_von_Hunden.

Cochrane, Alasdair. *An Introduction to Animals and Political Theory*. New York: Palgrave Macmillan, 2010.

Cooke, Steve. "The Ethics of Touch and the Importance of Nonhuman Relationships in Animal Agriculture." *Journal of Agricultural and Environmental Ethics* 34, no. 2 (2021): 1–20. DOI:https://doi.org/10.1007/s10806-021-09852-5.

Donaldson, Sue, and Will Kymlicka. *Zoopolis: Eine politische Theorie der Tierrechte*. Translated by Joachim Schulte. Berlin: Suhrkamp, 2013.

Francione, Gary L. *Animals, Property and the Law*. Philadelphia: Temple University Press, 2007.

Rechtsinformationssystem des Bundes (RIS). "Bundesrecht konsolidiert: Gesamte Rechtsvorschrift für 2. Tierhaltungsverordnung." Published n.d.-a. Accessed May 20, 2024. https://ris.bka.gv.at/GeltendeFassung.wxe?Abfrage=Bundesnormen&Gesetzesnummer=20003860.

Rechtsinformationssystem des Bundes (RIS). "Bundesrecht konsolidiert: Gesamte Rechtsvorschrift für Tierschutzgesetz." Published n.d.-b. Accessed May 20, 2024. https://ris.bka.gv.at/GeltendeFassung.wxe?Abfrage=Bundesnormen&Gesetzesnummer=20003541.

Rechtsinformationssystem des Bundes (RIS). "Bundesrecht konsolidiert: Gesamte Rechtsvorschrift für Tierversuchsgesetz 2012." Published n.d.-c. Accessed May 21, 2024. https://ris.bka.gv.at/GeltendeFassung.wxe?Abfrage=Bundesnormen&Gesetzesnummer=20008142.

Rechtsinformationssystem des Bundes (RIS). "Bundesrecht kondolidiert: Artikel § 285a." Published n.d.-d. Accessed April 22, 2016. https://www.ris.bka.gv.at/Dokument.wxe?Abfrage=Bundesnormen&Dokumentnummer=NOR12018011.

Rechtsinformationssystem des Bundes (RIS). "Bundesrecht konsolidiert: Artikel §§ 36 & 37. Published n.d.-e. Accessed November 26, 2015. https://ris.bka.gv.at/GeltendeFassung.wxe?Abfrage=Bundesnormen&GesGesetzesnum=20003541.

Tanasescu, Mihnea. (2014) "The Rights of Nature: Theory and Practice." In *Political Animals and Animal Politics*, edited by Marcel Wissenburg and David Schlosberg, 150–163. London: Palgrave Macmillan, 2014.

Wissenburg, Marcel. "An Agenda for Animal Political Theory." In *Political Animals and Animal Politics*, edited by Marcel Wissenburg, and David Schlosberg, 30-43. London: Palgrave Macmillan, 2014.

CHAPTER 14

Specific Rights in Nonhuman Animal Rights Declarations (Part 4)

14.1 Article 7: Right to Love, Sexuality, Procreation, Pair-Building, Family, and Friendships

> The point of love is always that it is also about our own happiness. (Precht 2009, 242)

> Love is Happiness. Full Stop. (cf. Vaillant 2012)

> Of course, in the end it is about love. […] Decisive for the success of a human life is love. (Sedmak 2012, 34)

> Whoever thinks that 'love' is a human achievement […] is mistaken. Love is one of the oldest inventions and an integral part of life. Love unites us, it does not separate us from our fellow creature, the animal. (Brensing 2013, 161)

> In the case of humans, the positive side of this claim [to bodily integrity] includes the opportunity for procreation and sexual gratification. […] All other things being equal, we should probably protect this capacity in animals as well. (Nussbaum 2010, 532)

D. Schneeberger, *Envisioning a Better Future for Nonhuman Animals*, The Palgrave Macmillan Animal Ethics Series, https://doi.org/10.1007/978-3-031-75332-9_14

It would be difficult to deny that the human needs for love, sexuality, procreation, family, and friendships tend to be strong needs. But how important is the fulfilment of these needs and the actualization of the corresponding abilities for nonhuman animal species? In order to be able to answer this question, we need to become more specific, and ask the question at least about each species; even then, there might be differences in the needs individuals within each species have. However, Nussbaum (2010, 535) argues for a nonhuman animal claim to loving attachments and relationships: "Animals have a complex emotional life. [...] Like humans, they are entitled to a life that enables them to form bonds with others, to meet them with love and concern, and not to be [...] denied these bonds." Since there are more findings about these needs in the human realm and these are generally conducive to a general understanding of the complex of phenomena surrounding love and sexuality in the nonhuman animal kingdom, findings with regard to these phenomena in the context of humans will also be discussed in this section (cf. epistemic anthropocentrism, Sect. 5.1 in this book).

The needs for love, sexuality, procreation and for a being socially involved are, on the one hand, primarily social needs (of humans, too) whilst on the other hand, the need for an individual to experience and actualize their own sexuality can also be understood as a more basic physical need (Maslow 1943, 381). In Maslow's hierarchy of needs, this group of needs is located partly in the first (physiological) level and partly in the third (social) level (ibid.).

The affirmation of life and the struggle for survival are programmed into us. We want to live at all costs (even if diseases and painful chronic conditions exist that may override this automatic urge to survive). Interestingly, it is not only our own survival that is of paramount importance to us, but also the continuation of our genes, our species. The biological reproductive programme conceals itself under the cloak of sexual desire, and among adult humans there are few who are never driven or motivated by it. Just as eating a tasty meal is accompanied by positive feelings, the reward centre in our brain becomes active during sexual activity: "It is generally accepted that activation of [...] [the] reward system is important for the formation of pair-bonds, and especially for making love a rewarding experience" (De Boer et al. 2012, 116).

Romantic love between two partners is understood in evolutionary and biological terms as a desired and genetically induced programme: “In evolutionary biology, love is seen as part of a long-term sexual strategy to maximize reproductive success. […] According to the Austrian behavioural scientist Irenäus Eibl-Eibesfeld (1928), the root of love lies in the mother-child bond, which developed in the course of intensive brood care in mammals” (Bliesener 2009, 345). It is not only mammals that form pair-bonds: many bird species and even certain species of fishes do too, including cichlids (Deutscher Tierschutzbund n.d.).

Sexual romantic relationships that endure over a period of time are biologically found most commonly where offspring require intensive care and nurturing, which in many cases can be better provided by two parents than by one: “In the case of a required stable partnership, the appropriate command is called: Love!” (Wohlleben 2014, 27). “A bonding mechanism in the brain ensures that individuals who have to perform important reproductive tasks together stay together and support each other” (Kotrschal 2013, 335).

Besides the pair-bonding common in various species, maternal love and care is similarly common. Caring, loving mothers who (aggressively) defend their offspring against possible attack can be observed in many species (including non-pair-forming ones): “Animal mothers would do anything for their offspring, even face their mortal enemies, just to know their child is well. There is nothing more dangerous in the wild than animal mothers who see no possibility of escape for the little ones” (Wohlleben 2014, 27).

Obviously, it is the case that the offspring also need this protection in order to survive and to grow up (and to sexually mature and then pass on their genes in turn). It is thus biologically wise to endow mothers with such love (including protective aggression against potential threats), if the goal of passing on genes is to be fulfilled, i.e. if evolutionary success is to occur; helpless newborns who are not lovingly cared for and protected from attackers have a difficult lot. This is not only the case with mammals; many young birds would starve if their parents were not constantly busy dropping food into their hungry mouths. For penguins who live in extremely cold climates, there would be no hatchlings if their parents did not constantly warm them under their bodies, for it only takes a short exposure to the cold air for budding life in the egg to be extinguished. The loving and reliable care of parents for their offspring is consequently a key biological programme.

Our emotions and sensations are an effective way of controlling us as sentient beings: "Emotions originate in the brain, in systems of neurons that we share almost unchanged with many other animals. Emotions drive behavior and are an important background for decision-making in humans and other animals" (Kotrschal 2013). We strive for pleasurable sensations and experiences and try to avoid painful and distressing ones. We seek the greatest possible happiness and pleasure, and find it (partly) in the arms of our partner and in sexuality: "Without positive feelings, there would be no sex and no reproduction" (Wohlleben 2014, 25). Hedonistic sexual beings are controlled and rewarded by the sweetest fruits of their brain chemistry and hormonal system: "In both love and drug use, humans strive to stimulate the brain's reward system" (Hurlemann 2013).

Loss of those we care about is painful. Events and conditions that endanger life and flourishing—our own or that of our children, or of beings close to us—are experienced as unpleasant and sorrowful: "[T]he loss of a parent can mean death to a young animal. Conversely, it throws a mother animal into emotional chaos when the child dies. Hinds leading a herd lose the leadership function after the loss of their calf. [...] The other herd members register that their boss is grieving, and they are not at all comfortable with it" (Wohlleben 2014, 29). Consider here, for example, the days-long crying of dairy cows for their calves. Hoy (2009, 83), cited in Sezgin (2014a, 156) reports on calf-weaning in dairy farming: "[B]oth calves and cows react with intense vocalization, increased locomotion, and reduced feed intake and resting times. On the first day after weaning, up to five hundred calls are emitted by the calf, and it takes days for this intense vocalization to subside." We (as well as other animal species) can scarcely envisage a worse psychological pain than the one we have to endure when we lose our children.

The harm many social beings suffer when important relationships they are in are ruptured are called "relational harms" (Cooke 2021). These kinds of harms have received little attention so far in the animal ethics debate (ibid., 3). Cooke (ibid., 2) points out that the "more intensive the farming, the more social the nonhuman animals farmed need to be." Relational harms are intrinsic in the animal industry (ibid., 3). "Farmed nonhuman animals have a strong interest in their relationships with one another, strong enough that they are seriously harmed by our controlling their lives" (ibid., 10).[1]

Various animal species have similar forms of relationships to the pair-formation that is common among Western humans. In the nonhuman animal kingdom, too, such relationships can be both heterosexual and homosexual in nature. Despite the fact that some animal species (e.g. various bird species) are monogamous, monogamy in mammals is generally the exception rather than the norm, as only about 3 to 5 per cent of mammal species exhibit such pair-bonding (Scheele et al. 2013, 20308). Sexual monogamy is a certain genetic cost factor for male individuals in terms of actually passing on genes, as they could spread them better by having sex with several female partners (ibid.).

At the Konrad Lorenz Research Centre in Grünau, Upper Austria, interesting observations have been made about the hetero- and homosexual pair-bonding of greylag geese:

> Kurt Kotrschal, head of the Konrad Lorenz Research Centre, likes to tell us what advantages same-sex goose couples enjoy in the social hierarchy: "As a single person, you are at the bottom of the gaggle of geese. As a bonded pair, you're in the middle, and as a bonded pair with offspring, you're at the top of the hierarchy." Thus, the evaluation of the life histories of 352 ganders

[1] In her book, *The Cow with Ear Tag #1389*, Kathryn Gillespie (2018, 196), reports a heartbreaking encounter she witnessed at an auction where cows and calves are sold:

> The first animal was already in the auction ring, ready to be sold: a day-old Holstein calf with a barcode sticker #604. The auction ring was clean: immaculate, really [...]. The tiny calf stood in the middle, looked around, and bleated. The calf's persistent cry was the only sound in the room, drowning out the distant noise of gates clanking and other animals bellowing from the pens outside. He was a new-born and his drying umbilical cord dangled from his belly as he stood there, slightly wobbly on his spindly legs. The man whose job it was to herd the animals through the auction pen entered the ring with a paddle and leaned on the fence in from of the auctioneer's platform to chat with the auctioneer. The calf, noticing the man, approached him and gently nudged his leg. He stood no taller than the man's knee and his nose nuzzled gently but persistently at the side of the man's thigh. In one efficient motion, the man turned and smacked the calf in the face with the paddle he was holding and spoke (much too loudly for the fairly quiet room), "I'm not your mother!" The calf leapt back, cowering, and then ran away from the man, across the pen.

This story vividly underscores the importance of empirical ethnographical work to shed light on the violent realities of the animal industry.

> showed that they have a good social status in a relationship with a male. Kotrschal calls this "homo-social behavior." (Schmidt 2008)

At the Konrad Lorenz Research Centre, it was also observed that the more often a heterosexual relationship has failed, the more likely it is that ganders will enter into a homosexual partnership:

> Geese can remain monogamous for life. But if one partner dies or the couple breaks up in a quarrel, the single individual has to look for someone else. After one failed relationship, 40 per cent of ganders enter into a "gay marriage," after the fifth relationship, as many as 80 per cent. Females have less pressure to find a partner because they live in sister clans that take care of each other. "True homosexual orientation, that is, ganders that never copulate with females in their entire lives, also exists". While the percentage is low, it shows unquestionably that same-sex inclinations are pre-programmed in nature. (Schmidt 2008)

Kotrschal (ibid.) emphasizes that monogamous partnerships can be profitable not only for the offspring, but also for the partners themselves. A stable pair-bond brings social benefits, as it is not only associated with increased social prestige, but also involves mutual caring (ibid.). Brensing (2013, 95) reports on the pair-bonding behaviour of dolphins: "The first pairs are observed at the age of six, and by the age of 20, over 90 percent of the animals have had a partner at some point. At the end of the dolphin's life, there are again many solitary individuals, indicating that the partner has died and that no new bond has been formed."

Just as with some species of fishes, marine mammals, and birds, prairie voles are also a pair-bonding species:

> The prairie vole is perhaps the most beautiful example of marital fidelity. The North American rodents celebrate the beginning of their relationship with a wild "night of love" lasting up to 24 hours. After that, they stay together for the rest of their lives, look after their offspring together and are even attached to their partner beyond death. When one of the two dies, the other no longer enters into a new relationship. (van Elderen 2013)

Oxytocin is the much-studied hormonal relationship glue that not only co-creates and holds together pair-bonds and close social bonds, but is also constitutively involved in feelings of wellbeing, happiness and security. Balluch (2014, 101) refers to research results (Julius et al. 2014) that show that considerable oxytocin releases are also involved as social glues in friendships between human and nonhuman animals and between individuals of different (nonhuman) animal species. Studies concerning human-dog interactions (Handlin et al. 2011; Miller et al. 2009; Nagasawa et al. 2009; Odendaal 2000; Odendaal and Meintjes 2003) consistently showed "that interaction with a dog is associated with an increased release of oxytocin in humans. This effect was particularly pronounced in interaction with one's own dog and in direct physical contact, indicating the importance of the emotional relationship between humans and dogs" (Julius et al. 2014, 81). Furthermore, an increased oxytocin level was also measured in the petted dog (Odendaal 2000; Odendaal and Meintjes 2003). Through these research results, Julius et al. (2014, 176) see the fact that domestic nonhuman animals such as dogs also develop affection and a social bond with humans as scientifically substantiated. This result also harmonizes with the tendency for nonhuman animals such as dogs, cats and horses to become increasingly attached to humans, to enter into friendly relationships with them and to live symbiotically with them (ibid.).

In another study (Miller et al. 2009), increased oxytocin levels could only be detected in female dog owners who interacted with their dogs. In female humans, the natural instincts of maternal love can thus presumably also be activated by nonhuman animals:

> In fact, the endocrine and physiological effects measured in breastfeeding mothers and their babies were similar to those observed in female dog owners and their dogs after these women petted their dogs. As with the mothers and their babies, oxytocin effects, such as low cortisol levels and falling blood pressure, were also observed in the women and their dogs. Handlin et al. (2011) conclude that these results should be interpreted as effects of the oxytocin system, as "part of the mammalian heritage," which "can be activated not only by individuals of the same species, but also by individuals of other species." (Julius et al. 2014, 178)

Humans also love their companion animals (like children) because they are beings who can accept and enjoy human care and reciprocate it. As stated above, loving interaction provides positive endocrine effects in the creatures involved, improving wellbeing and physical and mental health

(Julius et al. 2014, 181). Julius and colleagues (Julius et al. 2014, 182) list the positive effects of oxytocin that are crucial here: reduction of anxiety, stress, aggression and depression, stimulation and facilitation of social interaction and communication, increased trust in others, facilitated access to emotional states, and facilitation of learning processes.

For humans, relationships with nonhuman animals are thus also beneficial because they can be companion and care object in one and thus satisfy important human social needs (at least to a certain extent). Humans can rely on the stable, sincere friendship of their companion animal in emotionally difficult times. By caring for a beloved being, the caring human can obtain their "dose of oxytocin" and thus improve their own wellbeing (Julius et al. 2014, 183).

Nonhuman animals also show their affection and friendship towards humans through caring and friendship behaviour typical of their species. Balluch (2014, 77) describes Hiasl's species-typical social care behaviour and its importance:

> For Hiasl it is of great importance to groom each other. We are supposed to rummage through the body hair on our arms and head to find lice. They're not there, of course, but it's still a sign of bonding for him and strengthens our relationship. He enjoys grooming very much, smacks loudly and keeps it up for a very long time. Such intimacy between him and me always touches me deeply.

This example described by Balluch illustrates the value of interspecies friendships for nonhuman animals. So far, the importance of social bonds, be they family bonds or friendships, has not been discussed in depth when it comes to nonhuman animal rights. If the protection of these goods is to be taken seriously in the future, the social systems in which nonhumans live may not be disrupted so easily anymore and not without good reason. At a talk given in Vienna in 2023 at the Animal Morality Conference, animal scholar and ethicist Lori Gruen reminded us that we can only begin to understand the sociality of some nonhuman animals and the importance of their relationships for them. She emphasized the distress and suffering nonhuman beings can experience when they undergo the loss of those close to them. She stated that individuals might maybe be better conceptualized as fingers of a hand, and not as separate as we might think.

With regard to the complex of needs around love, sexuality and reproduction, there are parallels in the curtailment of these abilities and the prevention or obstruction of the fulfilment of the needs associated with them between disabled humans and nonhuman animals that (have to) live in multispecies societies with humans. In order to outline these parallels, and understand what can be learned from them for nonhuman animal rights, let us once again turn to insights from critical disability studies.

14.1.1 Right to Love, Sexuality, Procreation, Pair-Building, Family, and Friendships for Disabled Humans

Love and sexuality are biological functions that go hand in hand with biological abilities and needs. Nussbaum argues that living beings should be enabled as far as possible to exercise their natural capabilities and thereby be able to pursue their goals autonomously (Nussbaum 2010, 477). Therefore, a claim to be able to actualize these capabilities thus arises at least theoretically from the capability and need to love.

Needs and capabilities of love and sexuality span a wide range of possibilities of experience, sensation, and realization. Sexual capacities are listed under "bodily integrity" in Martha Nussbaum's extension of the capability approach to nonhuman animals. "Bodily integrity" commonly encompasses being unharmed physically and protection from unwanted interference with the body. Martha Nussbaum, for example, speaks out against female genital mutilation, which is still found in various cultures (cf. Galamaga 2014, 27). Circumcision of the sexual organs can have a substantial negative impact on the ability to feel and thus be detrimental to the valuable ability to experience sexual pleasure and satisfaction. Brian Earp (2014) argues against the unjustifiably much less stigmatized tradition of (involuntary) circumcision of male (among others, e.g. Jewish or US-American) infants. He (ibid.) explains—analogous to the discussion around female circumcision—that erogenous tissue, which can be important for pleasurable sexual feelings and fulfilling sexual experiences in later life, is irreversibly removed without the consent of the individual affected.

In the case of humans, the demand for a right to bodily integrity with regard to one's sexual organs and fertility goes along with a widespread intuition that humans have a right not to have their reproductive capacity (negatively) influenced against their will and that everyone should theoretically have the right to experience satisfying sexuality and to procreate. If sufficient care for the offspring cannot (yet) be ensured, it may follow that it is ethically desirable to prevent the creation of new life (at least until circumstances have changed in such a way that the newborn(s) can at least be sufficiently well cared for). A prerequisite for this norm, however, is easy access to reliable and (as far as possible) harmless contraceptive methods. Modern contraceptive methods free heterosexual sexual intercourse from a possible natural consequence, namely the creation of offspring. Thus, those beings who can be expected to deal responsibly with the creation of new life also have the responsibility to reproduce only to the extent that a good life can be expected for the resulting living beings and the necessary (financial, emotional and motivational) conditions for a dignified and happy life of the offspring are given.

The human right to marry and procreate is enshrined in the Universal Declaration of Human Rights in Article 3.1: "Men and women of marriageable age, without any limitation on the basis of race, nationality or religion, have the right to marry and to found a family." This Article thus gives rise to a right to voluntarily enter into partnerships and to voluntarily experience one's own reproductive capacity. Incidentally, according to the Universal Convention on the Rights of Persons with Disabilities (Article 23), persons with disabilities have the same right as non-disabled persons to enter into partnerships, to procreate and to found families.

Disabled and non-disabled humans have pronounced needs for partnership, sexuality and procreation: "For people with severe disabilities, living in a partnership (also a sexual one) is a need that, although often difficult to achieve, is nevertheless strongly desired and often persists into old age" (Rothaug 2013, 151). Rothaug cites the example of "Mr. K.," a severely disabled person, who is only able to move his head independently and communicates non-verbally via a speech computer, who formulates his situation, feelings, and wishes as follows: "I am, like most of us, probably, whether disabled or not, full of longing for human warmth, a firm hold, reliable relationships. [...] Love, closeness and the feeling of belonging to a person are the most important things in my life" (Rothaug 2013, 156).

Even in humans with disabilities, the desire for physicality, lived sexuality and offspring tends to be present and relatively urgent or pressing. This desire is a (or *the* primary) symptom of a natural biological programme that is responsible for a very important "interest of the genes": reproduction. Nevertheless, the realization of sexual needs and wishes to procreate are still often negatively influenced or hindered by caregivers: “For some people with learning disabilities, the unfulfilled desire to have children is harder to bear than the disability itself. In residential homes for disabled humans, there is sometimes still the “pill for everyone” so that pregnancy does not occur” (Göthling and Kühnel 2013, 106). In this context, Zinsmeister (2013, 63f) also refers to the fact that many residents of care facilities for the disabled are given a contraceptive injection every three months, regardless of the concrete risk of pregnancy. One reason for this, Zinsmeister (ibid.) assumes, is the fact that this hormonal injection prevents menstruation, which makes work easier for the nursing staff; however, as Zinsmeister (ibid.) also points out, the administration of this injection is potentially harmful to the health of the users (especially if used for years).

Cognitively disabled humans are sometimes manipulated by their environment into excluding and ignoring their (possible) desires for sexuality and procreation (Herrath 2013). Sterilizations are also caused by such influence: “Many parents—especially of daughters—fear that their daughter will have a child and be overwhelmed by it” (Pro Familia 2006, 15). Women with learning difficulties were and are therefore often urged by parents or carers to have a sterilization” (Arnade 2013, 41).

The extent to which such paternalistic interventions are sufficiently justified needs to be examined critically. It is undoubtedly the case that these interventions, or the prevention of producing offspring, can be justified in certain cases. Nevertheless, it must be taken into account which abilities, possibilities and freedoms disabled humans (in particular, according to the Universal Convention on the Rights of Persons with Disabilities [Article 23]) are justly entitled to, and which paternalistic interventions violate the rights of humans with disabilities.

When we think of the right to (sexual) love, we probably first think of a negative right, i.e. that living beings should have the freedom to pursue this need independently and to seek fulfilment in encounters, interactions, and relationships based on mutual consent between adults in a self-determined way, without being hindered or negatively influenced by anyone. It is therefore not a positive right: i.e. there is no claim that living beings must or should be helped from the outside to fulfil this need.

However, a negative right can in certain cases be complemented by a positive right. A positive right to the possibility of actualizing their own sexuality for an individual might include a right to the provision of sexual assistance or sexual services (if desired by the individual concerned): “It is important to distinguish between passive and active assistance. While passive assistance can be the provision of aids, creating a nice environment, leaving out the incontinence pad or much more, active assistance is reserved for sexual assistants coming from outside" (Rothaug 2013, 153).

In modern welfare states it is also the case that women with children or families have a right to support from the state, for example, in the form of child benefits, entitlement to parental leave, protection against dismissal during pregnancy, free kindergarten, tax benefits, etc. Article 23 of the Convention on the Rights of Persons with Disabilities is (as is usual for human rights) vague and thus it is not entirely clear which obligations *exactly* arise from it (my emphasis):

> Article 23
> Respect for the home and the family
>
> (1) States Parties shall take effective and appropriate measures to eliminate discrimination against persons with disabilities on an equal basis with others in all matters relating to marriage, family, parenthood and partnerships to ensure that
> (a) The right of all persons with disabilities of marriageable age to marry and to found a family on the basis of the free and full consent of the future spouses is recognised;
> (b) The right of persons with disabilities to *decide freely and responsibly on the number of children they have and the spacing of births*, and to have access to age-appropriate information and education on reproduction and family planning, is recognised and they are provided with *the necessary means to exercise these rights*;
> (c) People with disabilities, including children, retain their fertility on an equal basis with others.
> (2) States Parties shall ensure the rights and responsibilities of persons with disabilities in matters of guardianship, conservatorship, custody of persons and property, adoption of children or similar legal institutions, to the extent that national law recognizes such; *in all cases, the best interests of the child shall be the determining factor. States Parties shall provide appropriate support to persons with disabilities in the exercise of their parental responsibilities.*

The individual's right to actualize their own sexuality is not explicitly mentioned here. This is most likely due to the fact that in formulating this right, care was taken to find a wording that would be agreed to by as many nations as possible: "[It] is not without a cause that the Convention on the Rights of Persons with Disabilities itself does not contain any explicit statement on sexuality—it is not only Islamist-oriented governments that are aversively positioned in this regard" (Herrath 2013, 29). Rights to partnership, fertility and offspring in some way, however, imply a right to a sexual life. It is thus imperative that persons with disabilities receive support in exercising their rights to partnership, sexuality, family and offspring. This is clearly a positive obligation. However, what form this support must or should take and how far it must extend is still open to question. For example, an entitlement to a 24-hour caregiver or supervision for mothers and their children could also be a possible reading of this text—in the case of certain impairments of the mother (or parents), the best interests of the child might possibly only be guaranteed sufficiently with such intensive support. Moreover, such support would probably have to take place over many years. This could accumulate costs for the state that would have to be financed socio-politically.

When support is referred to as "reasonable," it can probably be assumed that the necessary support is at least provided. Since having a carer present around the clock for each family would usually be difficult to finance, it is more likely that, in reality, disabled parents would have their right of care withdrawn, were they unable to fulfil the duty of care (independently) sufficiently well. Children of humans with an impairment that negatively affects their ability to care may be placed in foster families or children's homes by the youth welfare office if there are no other humans in the immediate family environment, such as grandparents, who could and would take over the duty of care for the children. Considering these problems in practice, it becomes easier to understand why the fertility and reproduction of humans with (mental) impairments tend to be curtailed by their environment. However, due to modern reliable contraceptive methods, heterosexual intercourse is no longer necessarily linked with procreation. Thus, enabling humans to actualize their sexuality and fulfil their sexual needs—at least in the sense of a negative right—would be necessary from an ethical point of view in order not to further negatively influence the quality of life and the subjectively perceived happiness in life of the people concerned: "If [...] adult people cannot have experiences of tenderness, security and physical closeness, something like a second disability often arises" (Göthling & Kühnel 2013, 105).

14.1.2 *Right to Love, Sexuality, Procreation, Pair-Building, Family, and Friendships for Nonhuman Animals*

Let us now focus once more on nonhuman animals. Many nonhuman (social) animals are also beings with (pronounced) sexual needs and strong needs for (active and passive) affection, social closeness, and love. The neurological systems and the neurotransmitters involved that bring about the neurochemical reward for these activities are equivalent in many nonhuman animal species to those effective in human neural circuits. If a sexual being is able to actualize its sexual abilities and needs, it experiences satisfaction (and thus contentment and wellbeing); if this is not the case, sooner or later frustration may occur, which is commonly perceived as unpleasant; in accumulated form, it can be detrimental to the health and flourishing of a being (resulting in depression, aggression, and stress).

Animals of many species go to considerable lengths to ensure their individual success in finding a mate. Great efforts might go together with great values or great importance; consider, for example, the self-presentation of male ducks: “At mating time, males perform ‘duck dances’ to impress females and secure a permanent mate” (Albert Schweitzer Stiftung n.d.). Geese are also partnership- and family-oriented:

> Geese are very loyal to their families: they choose a mate for life and protect their mate and children. If a goose gets sick or injured, the mother or mate often refuses to leave her side, even when winter approaches and the other geese in her group fly south. (People for the Ethical Treatment of Animals (PETA) n.d.)

Brensing (2013, 91) points out that dolphins become highly active when looking for partners and their actions revolve around achieving this goal in certain phases of their lives. Brensing (ibid.) compares the situation with human teenagers or young adults, who are hormonally programmed for the goal of reproduction, and in whom the desire and search for a partner and a stable sexual relationship also automatically germinates sooner or later (ibid.).

> The interesting thing about this thought is actually the fact that in the situations that really mean something to us, that move us emotionally a lot and give meaning to our lives, we are no different from many other highly developed animals living in groups. It is more a question of how we look at the

situation, whether we behave animal-like or whether animals behave human-like. The common denominator, our inner motivation has evolved and helps us to preserve our species.

It would be rather strange if those very phenomena that humans might experience as particularly "animal-like" (see sexuality) were to distinguish them from other animals. It is probably more likely that the phenomena of love, sexuality, pair-bonding and reproduction and most of the associated needs, interests and abilities are, as Brensing also states, evolutionary commonalities. It should perhaps also be assumed that it is not only the case with us humans that relationships with romantic partners and family (apart perhaps from our own existence) tend to be the most important goods in life. Obviously, these relationships are precious, not only for human animals:

> For me, as a behavioral biologist, love is one of the most primitive biological mechanisms [...]. Since the invention of the two sexes, nature has had to come up with a mechanism to reunite them. We have many names for it: Sex drive, partnership or even love. [...] Because this mechanism has worked so well, Mother Nature has developed it further in the course of evolution. Therefore, many animals [...] "love" their offspring, often more than their own lives. (Brensing 2013, 160f)

"Mother Nature" cleverly arranges that parents protect their beloved offspring with their lives under certain circumstances, should this be necessary. For as far as the passing on of genes is concerned, the offspring might be the more important individual, the individual who has the potential to help the genes advance further into the future and to ensure their continuation in further offspring, whose life can once again advance further into the future. Thus the "immortality urge" of our genes is met and satisfied.

There are humans who (for whatever reason) spend their lives without a partner, without experiencing shared sexuality and family with another being in a satisfying way. Is experiencing the goods of partnership, (satisfying and fruitful) shared sexuality and family of such importance to a human life that their absence equals an impoverished human life? Given the importance attached to these facets of (human) life, this is open to debate. This importance is attributed to the presence of these phenomena in particular because of the happiness that these goods represent and the fundamental sadness that individuals who do not find love might experience. On the one hand, the subjectively felt mental strain that accompanies a frustration of the need for love, partnership, and sex arises from the nature of the being; it tries to drive the individual into the fulfilment of the genetic mandate through unpleasant sensations. It is not only humans who are driven by the urgent desire to have the goods offered by partnership, sexuality, and reproduction. After all, it is about biological mechanisms that aim at the successful transmission of genes. It is about life that wants to live, and not just until an individual dies, but beyond.

Currently, the sexuality and fertility of nonhuman animals is controlled, influenced and destroyed (sometimes radically, and on a large scale) by humans. We castrate them without duly considering the value of the destroyed reproductive capacity or the (inherent and intrinsic) value of a sexual life and fertility for nonhuman animals. The essential damage that sexual individuals experience as a result is usually largely ignored by us; we often even assume—perhaps sometimes correctly, perhaps sometimes wrongly—that it is in the interest of the nonhuman animal (or its potential, putative offspring) to be castrated.

As already mentioned in Chap. 11, in Austria, cat "owners" whose cats have outdoor access are obliged by law to have them neutered (Rechtsinformationssystem des Bundes (RIS) n.d.). The aim of this law is to redress the imbalance between existing living options in human households and the number of domesticated cats, as domesticated cats tend to need to live in human homes in order to thrive. If fewer kittens are born, fewer unwanted nonhuman animals who do not find a place in an interspecies society with humans will experience suffering and be killed. Sezgin (2014a, 293) advocates for neutering domesticated nonhuman animals when there are good reasons to do so:

> I have no fundamental objections to the castration of domesticated animals. Certainly, here we are curtailing an important component of the animal's life (which, by the way, we are not so much depriving of sex, which for most nonhuman animals becomes important only in circumscribed rutting seasons, but rather of the possibility of rearing offspring). [...] [It is clearly better] to deprive mothers of the right to have offspring than to deprive their children of life.

With regard to human individuals, we share the intuition that it is ethically reprehensible to interfere with their fertility without their consent (or against their will), thereby depriving them of something valuable or priceless. In comparison, it is common practice to intervene (irreversibly) in the sexuality and reproduction of nonhuman animals at the discretion of humans. Why are the fertility and sexuality of humans so important that no one is allowed to negatively interfere with them without consent, but those of nonhuman animals are not? For nonhuman animals, just as with humans, sexuality, reproduction, and pair-bonding can represent a significant potential source of intrinsically valuable and (telos-)fulfilling capabilities, experiences and activities. Many nonhuman animals are also deeply sexual beings (like humans). We humans take this away from nonhuman animals without much thought. Is this really sufficiently justified?

As already mentioned in the discussion of the concept of telos and repeated here for completeness, Korsgaard's (2011, 110) neo-Kantian animal ethics imperative is suitable for gaining insights into the right to sexuality and fertility of nonhuman animals. According to this maxim, we should interact with nonhuman animals in ways that we could plausibly assume they would consent to if they could. It is reasonable to assume that nonhuman animals would also advocate for a right to bodily integrity, including with respect to their sexual organs and their natural functionality, and, in fact, they do so, for example, by automatically trying to escape painful procedures (such as castration) by fleeing or defending themselves ("voting with their feet," one might say).

We need to consider what accounts for this discrepancy in our views and actions between human and nonhuman animals when it comes to the right to sexuality, procreation, partnership and family. Could it be that we deny nonhuman animals their right to these goods because we do not understand, or have not yet understood, that for them, just as for us humans, these are essential aspects of their special and individual telos and, in this sense, part of the essential experience of their lives as sexual, social, naturally loving beings, and, on the other hand, in a certain sense, a measure of their quality of life and happiness in life? Likely we are not concerned with their telos, because we are egoistically used to thinking about ours, and barely take theirs into account: "When human activity pre-empts or interrupts the life project or teloi of nonhuman animals or ecological systems, the central issue is the lack of recognition of the validity and integrity of the natural world" (Schlosberg 2014, 89ff):

> About half of my sheep, some male, some female, have been full-grown for years but still like to stay close to their mothers. Daughters and mothers in particular go out to pasture and sleep next to each other; Jane, for example, who has long since grown up, likes to lay her big head on mum's back while chewing the cud. In humans, this is called a family. (Sezgin 2014b, 23)

Donaldson and Kymlicka (2013, 324; my emphasis) advocate for some demanding normative implications regarding the phenomena discussed in this section: "Once domesticated animals are recognised as citizens, they have rights, which include the *right not to have their sexual and reproductive activities unnecessarily curtailed*, as well as the right to have their offspring cared for by the mixed human-animal society." From the latter entitlement to assistance in caring for their offspring, Donaldson & Kymlicka (ibid.) conclude that some restriction on the actual reproduction of nonhuman animal citizens by humans in a zoopolis may be justified on practical (financial) grounds. Donaldson & Kymlicka (ibid.) demand that these justified interventions involve as little invasiveness, damage and restriction as possible: suitable methods would include contraceptive vaccines, temporary separation or non-fertilization (in the case of chicken eggs). For Donaldson and Kymlicka (ibid., 320), such human interventions are justified, since they are comparable to the demand on humans not to reproduce if adequate support for offspring cannot be guaranteed. For Donaldson and Kymlicka (ibid., 319), sterilization and castration are accompanied by (considerable) violation of the rights of a nonhuman animal citizen, and thus require a convincing justification.

Furthermore, Donaldson and Kymlicka (ibid., 324f; my emphasis) come to the following conclusion regarding changes to be sought in the area of breeding nonhuman animals: "Instead of the situation prevailing today—in which only some domesticated animals are declared breeding animals, while the overwhelming majority do not reproduce at all—*the chance to have and raise young could be granted to the majority of animals*, but at the same time *the overall scope for reproduction could be restricted*." This proposal does justice to both the telos and needs of a sexual living being and the resulting sexual and reproductive rights of that individual. If more nonhuman animals were granted a right to sexuality and family formation, including the exercise of care for offspring, more nonhuman animals kept by humans would have the chance a life that includes more aspects of their telos and the fulfilment of their needs and actualization of their capabilities. This exercise of capabilities and the fulfilment of sexual and reproductive needs corresponds to an enrichment of a nonhuman animal life and, due to its more manifold (potentially at least temporarily) gratifying of perceived possibilities of experience, to an increased or enlarged intrinsic value. In addition, the fact that humans have damaged many animal species through breeding (ibid., 327) speaks in favour of leaving this path, and returning to natural diversity. Genetic diversity within a species is, after all, a natural biological protective mechanism to ensure that a species thrives, since genetically divergent populations of a species or group of animals are more adaptable and resilient. All in all, one could conclude concerning these rights that a way forward is to enable/ not hinder nonhuman beings as far as possible, and to interfere with the least harmful method where absolutely necessary.

In the following chapter, I discuss challenges and ethical considerations surrounding property rights for nonhuman animals. Conservationists, environmentalists, and animal rights activists could join forces against destructive human interventions that endanger vital habitats for nonhuman fauna. The chapter argues that the right to life inherently includes the right to a suitable habitat. I again refer extensively to Donaldson and Kymlicka, who make the case that nonhuman animals require protection for their right to live autonomously in their territories, emphasizing their sovereignty. Critiquing traditional anthropocentric property theories that exclude nonhuman animals based on personhood criteria, the chapter proposes alternative justifications rooted in the basic needs of nonhuman animals for their territories. It introduces concepts such as "home range" and "core area" as alternatives to traditional definitions of the hard-to-grasp concept of "territory." The chapter explores legal and ethical challenges in attributing property rights to nonhuman animals.

The intuition that it is morally praiseworthy to protect the habitat of nonhuman animals living in the wild is prominent in the discussion about a right to property. It is, however, not shared by all the scholars in the field. Some scholars in the wild animal suffering discourse argue that instead of protecting habitats, we should cover natural areas with concrete to extinguish all life there. It is this counter-intuitive controversial discussion which will be dealt with in the discussion of the wild animal suffering discourse in the section after the next one.

References

Albert Schweitzer Stiftung. "Die Tiere: Enten." Published n.d.. Accessed November 7, 2015. albert-schweitzer-stiftung.de/massentierhaltung/enten.

Arnade, Sigrid. "Sichtbarer den je: Würde und Chancengleichheit – Die Behindertenrechtskonvention und die sexuelle Selbstbestimmung behinderter Mensch." In *Sexualität leben ohne Behinderung: Das Menschenrecht auf sexuelle Selbstbestimmung*, edited by Jens Clausen and Frank Herrath, 35–46. Frankfurt: Kohlhammer, 2013.

Balluch, Martin. *Der Hund und sein Philosoph: Plädoyer für Autonomie und Tierrechte*. Vienna: Promedia, 2014.

Bliesener, Thomas. *Brockhaus Psychologie – Fühlen, Denken und Verhalten verstehen*. Leipzig: F.A. Brockhaus, 2009.

Brensing, Karsten. *Persönlichkeitsrechte für Tiere*. Freiburg im Breisgau: Herder, 2013.

Cooke, Steve. "The Ethics of Touch and the Importance of Nonhuman Relationships in Animal Agriculture." *Journal of Agricultural and Environ mental Ethics* 34, no. 2 (2021): 1–20. DOI:https://doi.org/10.1007/s10806-021-09852-5.

de Boer, A., E. M. van Buel, and G. J. ter Horst. "Love is More Than Just a Kiss: A Neurobiological Perspective on Love and Affection." *Neuroscience* 201 (2012): 114–124.

Donaldson, Sue, and Will Kymlicka. *Zoopolis: Eine politische Theorie der Tierrechte*. Translated by Joachim Schulte. Berlin: Suhrkamp, 2013.

Earp, Brian. D. "Female genital mutilation (FGM) and male circumcision: Should there be a separate ethical discourse?" *Practical Ethics*. Oxford: University of Oxford, 2014. https://www.academia.edu/8817976/Female_genital_mutilation_FGM_and_male_circumcision_Should_there_be_a_separate_ethical_discourse. DOI: https://doi.org/10.13140/2.1.3530.4967.

Galamaga, Adam. *Philosophie der Menschenrechte von Martha C. Nussbaum: Eine Einführung in den Capabilities Approach*. Marburg: Tectum Verlag, 2014.

Gillespie, Kathryn. *The Cow with Ear Tag #1389*. Chicago: University of Chicago Press, 2018.

Göthling, Stefan, and Anita Kühnel. "Wer Sexualität erleben möchte, sollte dafür Raum zur Verfügung haben' – Über sexuelle Stellvertretung und sexbezogene Behindertenhilfe." In *Sexualität leben ohne Behinderung: Das Menschenrecht auf sexuelle Selbstbestimmung*, edited by Jens Clausen and Frank Herrath, 102–107. Stuttgart: Kohlhammer, 2013.

Handlin, Linda, Eva Hydbring-Sandberg, Anne Nilsson, Mikael Ejdebäck, Anna Jansson, and Kerstin Unväs-Moberg. "Short-term interaction between dogs and their owners – effects on oxytocin, cortisol, insulin and heart rate: an explanatory study." *Athrozoös* 24 (2011): 301–316. DOI: https://doi.org/10.2752/175303711X13045914865385.

Herrath, Frank. "Human Right Meets Life Reality: What Disables Sexuality?" In *Sexualität leben ohne Behinderung: Das Menschenrecht auf sexuelle Selbstbestimmung*, edited by Jens Clausen, and Frank Herrath, 19–34. Stuttgart: Kohlhammer, 2013.

Hoy, Steffen, ed. *Nutztierethologie.* Stuttgart: Verlag Eugen Ulmer, 2009.

Hurlemann, René. Cited November 27, 2013. In "Forscher der Ruhr-Uni Bochum entdecken Hormon, das Männer treu macht." Accessed Febraury 21, 2015. derwesten.de/staedte/bochum/forscher-der-ruhr-uni-bochum-entdecken-hormon-das-maenner-treu-macht-id8709690.

Julius, Henri, Andrea Beetz, Kurt Kortschal, Dennis C. Turner, and Kerstin Unväs-Moberg. *Bonding with animals: Psychological and neurobiological foundations of animal-assisted interventions.* Göttingen: Hogrefe Verlag, 2014.

Korsgaard, Christine M. "Interacting with Animals: A Kantian Account." In *The Oxford Handbook of Animal Ethics*, edited by Tom L. Beauchamp, and Raymond G. Frey, 91–118. Oxford: Oxford University Press, 2011.

Kotrschal, Kurt. "Warum Menschen mit anderen Tieren sozial sein wollen und können." In *Tiere: Der Mensch und seine Natur*, edited by Konrad Paul Liessmann, 333–359. Vienna: Paul Zsolnay, 2013.

Maslow, Abraham. "A Theory of Human Motivation." *Psychological Review* 50, no. 4 (1943): 370–391. DOI: https://doi.org/10.1037/h0054346.

Miller, Suzanne C., Cathy Kennedy, Dale DeVoe, Matthew Hickey, Tracy Nelson, and Lori Kogan. "An examination of changes in oxytocin levels in men and women before and after interaction with a bonded dog." *Anthrozoös* 22 (2009): 31–42. DOI: https://doi.org/10.2752/175303708X390455.

Nagasawa, Miho, Takefumi Kikusui, Tatsushi Onaka, and Mitsuaki Ohta. "Dog's gaze at its owner increases owner's urinary oxytocin during social interaction." *Hormones and Behavior* 55 (2009): 434–441. DOI: https://doi.org/10.1016/j.yhbeh.2008.12.002.

Nussbaum, Martha C. *Die Grenzen der Gerechtigkeit: Behinderung, Nationalität und Spezieszuhörigkeit.* Translated by Robin Celikates and Eva Engels. Frankfurt am Main: Suhrkamp, 2010.

Odendaal, J.S. "Animal-assisted therapy: magic or medicine?" *Journal of Psychosymatic Research* 49 (2000): 275–280. DOI: https://doi.org/10.1016/s0022-3999(00)00183-5.

Odendaal, J.S., and R.A. Meintjes. "Neurophysiological correlates of affiliative behaviour between humans and dogs." *Veterinary Journal* 165 (2003) 296–301. DOI: https://doi.org/10.1016/s1090-0233(02)00237-x

People for the Ethical Treatment of Animals (PETA). "Entenmast." Last modified n.d.. Accessed November 7, 2015. peta.de/entenmast#.Vj5blpUrmUk.

Precht, Richard David. *Liebe – ein unordentliches Gefühl.* Munich: Goldmann, 2009.

Pro Familia. *Sexualität und geistige Behinderung.* Frankfurt am Main: Pro Familia, 2006.

Rechtsinformationssystem des Bundes (RIS). "Bundesrecht konsolidiert: Gesamte Rechtsvorschrift für 2. Tierhaltungsverordnung." Published n.d.. Accessed May 20, 2024. https://ris.bka.gv.at/GeltendeFassung.wxe?Abfrage=Bundesnormen&Gesetzesnummer=20003860.

Rothaug, Martin. "Sexuelle Selbstbestimmung bei schwerer Behinderung." In *Sexualität leben ohne Behinderung: Das Menschenrecht auf sexuelle Selbstbestimmung*, edited by Jens Clausen, and Frank Herrath, 47–71. Stuttgart: Kohlhammer, 2013.

Scheele, Dirk, Andrea Wille, Keith M. Kendrick, Birgit Stoffel-Wagner, Benjamin Becker, Onur Güntürkün, Wolfgang Maier, and Rene Hurleman. "Oxytocin enhances brain reward system responses in men viewing the face of their female partner." *Proceedings of the National Academy of Sciences of the United States of America* 110.50 (2013): 20308–20313. DOI: https://doi.org/10.1073/pnas.1314190110.

Schlosberg, David. "Ecological Justice for the Anthropocene." In *Political Animals and Animal Politics*, edited by Marcel Wissenburg, and David Schlosberg, 75–89. London: Palgrave Macmillan, 2014.

Schmidt, Veronika. "Ganz normal: 'Homo-Ehe' im Tierreich." Published April 15, 2008. Accessed December 27, 2014. diepresse.com/home/science/376626/Ganz-normal_HomoEhe-im-Tierreich.

Sedmak, Clemens. *Geglücktes Leben*. Freiburg: Herder, 2012.

Sezgin, Hilal. *Artgerecht ist nur die Freiheit*. Munich: C.H. Beck, 2014a.

Sezgin, Hilal. *Hilal Sezgins Tierleben: Von Schweinen und anderen Zeitgenossen*. Munich: C.H. Beck, 2014b.

Vaillant, George E. "Happiness is Love: Full Stop". Unpublished manuscript, 2012. Accessed November 20, 2024. https://www.duodecim.fi/xmedia/duo/pilli/duo99210x.pdf.

van Elderen, Heidi. "Bei manchen Tieren hält die Liebe ein Leben lang." Last modified February 14, 2013. Accessed December 27, 2014. tierwelt.ch/?rub=4503&id=34463.

Wohlleben, Peter. *Die Gefühle der Tiere*. Darmstadt: pala-verlag, 2014.

Zinsmeister, Julia. "Rechtsfragen der Sexualität, Partnerschaft und Familienplanung." In *Sexualität leben ohne Behinderung: Das Menschenrecht auf sexuelle Selbstbestimmung*, edited by Jens Clause, and Frank Herrath, 147–159. Stuttgart: Kohlhammer, 2013.

CHAPTER 15

Specific Rights in Nonhuman Animal Rights Declarations (Part 5)

15.1 Article 8: Right to Property: Ownership, Stewardship, or Burning Down the Forest?

Due to the growth of human populations and the steady spread of humans, the habitats of more and more nonhuman animals living in the wilderness are threatened. Consequently, nature conservationists, environmentalists and animal rights activists—despite differing in their ethical positions and values—can join forces against disruptive human interventions or the (total) destruction of habitats. A right to life must also imply a right to suitable habitat; where there is no habitat, there is no life, or as Bradshaw (2020, 32) puts it: "[C]learing a forest to make a field has the same effect on wildlife that a slaughterhouse has for livestock." Nonhuman animals also defend their territories in various ways; there are sometimes bitter territorial fights between individual nonhuman animals or between groups of them. Donaldson and Kymlicka (2013, 394; my emphasis) explain, with regard to nonhuman animals living in the wild: "What they need is not (or not only) a property right to their special nest or to their own den, but they need protection of their right to continue to live their own lives in *their own territory*. In short, what they need is *sovereignty*." Donaldson and Kymlicka (ibid., 374) further suggest that injustices between humans and nonhuman animals living in the wild should be addressed "by setting

D. Schneeberger, *Envisioning a Better Future for Nonhuman Animals*, The Palgrave Macmillan Animal Ethics Series,
https://doi.org/10.1007/978-3-031-75332-9_15

fair terms for interactions between sovereign communities." Nussbaum (2010, 538) states that the human right to property corresponds to a nonhuman animal right to habitat. It follows that humans should respect the territorial integrity of the habitat of nonhuman animals.

Territorial behaviour is indicative of a need for one's own territory. It makes sense to defend something that is of value. A territory provides access to natural goods needed for the satisfaction of basic needs (Hadley 2015, 15). When it comes to (the notion of) nonhuman animal property, territorial behaviour plays a central role (ibid., 2). Both human and nonhuman animals mark and defend the boundaries of their territory. It is interesting to note that occupying a territory—which brings with it a sense of "this (land) is mine"—is not a human institution: rather, it is an institution we share with many nonhuman species. Nonhuman animals "can create, follow, and enforce property rules among members of conspecies and even in some interspecies disputes" (Bradshaw 2020, 72). However, due to the power imbalance with humans, nonhuman animals "are forever in a state of vulnerability" (ibid.) when it comes to their property rights. Bradshaw (ibid., 23) points out that some nonhuman animals use sounds or smells to demarcate their territory. She (ibid.) also states that dogs are excellent boundary defenders of the places they occupy, also when they live together with humans. Nonhuman animals thus have strong property interests; we have just failed to understand them (ibid., 6) and take them seriously. The territorial behaviour that nonhuman animals exhibit can be used as a way of determining the boundaries of their territory (Hadley 2015, 24). What counts as the territory of a nonhuman animal individual or group can, however, be hard to decipher. Some questions arise: "Must an animal reside in its territory or is it sufficient just to pass through? If passing through marks out a territory, how often must an animal pass through: every day, once a week, twice a month? [...] [M]ust the animals do something species-specific, say, scent mark, give birth, or engage in defensive behaviour?" (ibid., 25) Due to such struggles to define territories, some ecologists opt to avoid the use of the territory concept and use concepts such as "home range" or "core area" instead (ibid.). These concepts refer to areas of land in which a nonhuman animal has an interest (ibid.). Home ranges and core areas are the "sites at which the animal engages in its most important species-typical behaviors: foraging, mating, and warding off challenges from conspecifics" (ibid., 26).

So, in order to theoretically advocate for the protection of the habitat of a group of nonhuman animals, one could resort to the concept of property. When we examine the current legal system, we see that the concept of property brings with it its own theory, which is again intermingled with other theories, namely the theory of transfer and the theory of justice in holdings (Hadley 2015, 40). Theories of property must entail information about how property can be acquired, how it can be transferred, and "what owners must do to ensure their property is secure" (ibid.). The Lockean labour theory of property is one of the leading theories when it comes to how ownership can be acquired (ibid.). It states that individuals who have "mixed their labour" with the existing goods at a certain place become owners of it (ibid.). Hadley (ibid.) cites an example proposed by James Rachels (1976), comparing a squirrel, who labours to gathers nuts for the upcoming winter, to human labour, pointing out that there is no relevant difference in how this could ground ownership of a territory. Another justification of how a certain part of land can become acquired is first occupancy (ibid., 50). Hadley points out here that at the centre of those theories is the idea that only human persons can own land: the general assumption that only "persons" (i.e. humans) can own land is, according to Bradshaw (2020, 1), a key driver of biodiversity loss.

In order to meet the criteria to be a person, individuals would, for example, have to be autonomous and self-conscious (Hadley 2015, 41). Some might argue that certain nonhuman animals might not meet those criteria. Some might claim that they are not persons and are thus unable to own land. Due to this person-centredness of the property theory, it "is very difficult to give animals property rights on labour or first occupancy grounds" (Hadley 2015, 50). Hadley (ibid., 51) continues: "If a person-centred basis for animal property rights is not a theoretically viable option, an alternative normative grounding must be found." Maybe the terms "animalhood" (Castelló 2022) and/or "beingness" (Deckha 2021) might prove helpful here (see Chap. 12).

Alongside the labour and the first occupancy justification, Hadley (2015, 55) discusses a potentially strong justification for why beings should have a right to their territory: they need it to fulfil their basic needs. This justification does not demand nonhuman animals be persons in the first place; instead: "[a]ll that needs to be accepted is the relatively modest claim that animals have an experiential welfare in the sense that their lives can go better or worse for them" (ibid., 56). Hadley (ibid., 59f) then discusses the "multiple-owner problem." He (ibid.) points out that different beings will have a property right on the same land and suggests that nonhuman animal property rights could be administered as group-rights assigned to the beings living in one area.

Owners are—to a certain extant—legally protected from—certain kinds of—interference with their property. Easements are legal tools that restrict an owner's power in certain ways. Hadley provides the examples of a right of way, or a right to fish (ibid., 64). Easements provide nuance to the ownership concept, which could otherwise very much be an all-or-nothing/black-and-white concept. Tanasescu (2014, 160) points out that we need to think carefully about whether we actually give wild animals their habitats as their property, especially given the consequences that attribution of ownership might bring. She (ibid.) outlines an example mentioned by Christopher Stone (2010, 169):

> [B]y strictly protecting owls (thus giving them de facto rights under the Endangered Species Act), the forest in which they live becomes the owls' property. Society as a whole might value the timber of some forest acreage more highly than it values the owls that depend on it. But once the owls are "listed," the owls prevail. And note that [the owls'] habitat is protected (as are our homes and lawns) from having *modifications* imposed upon them.

Hadley (2015, 20), however, points out that the concept of ownership provides scope for nuance: "it's just wrong to think of owners as having absolute discretion about what they do with their property."

Along with providing this nuanced understanding of ownership, Hadley (ibid., 69) points to another roadblock which might potentially obstruct nonhuman animal property rights: Can property own property? Hadley (ibid.) states that the legal status of "wild animals" "is an open question" (i.e. they are no one's property). It could therefore be logically relatively straightforward to give property rights to nonhuman animals living in the wild (ibid.).

Ownership can also be understood as a form of stewardship (ibid.). Perhaps attribution of ownership is not the most appropriate conceptualization when it comes to protecting wildlife habitats. Protection of wildlife habitats could alternatively be achieved through a stewardship model: "According to this model, wildlife habitats are protected in the form of protected areas, reserves or national parks. These wildlife areas are subject to management or guardianship by humans and are intended to serve the common good of humans and wildlife" (Donaldson and Kymlicka 2013, 375). Donaldson and Kymlicka (ibid., 376) take issue with this variant of habitat protection in that nonhuman animals are conceptualized as "incompetent beings" and "passive addressees of our actions" (ibid.). In contrast to the stewardship model, the sovereignty model (or "wild animal sovereignty theory") proposed by Donaldson and Kymlicka does not involve the right of humans to "govern" the habitat or to make caretaker decisions: "When and if humans visit the territory of wild animals, they do not appear there as caretakers and managers, but as visitors to an alien land" (ibid.). This second model corresponds most closely to a hands-off approach. Hadley (2015, 78) criticizes this theory, calling it "a misrepresentation of animal rights theory to suggest that it is a strict negative rights theory." He maintains that a provision of goods to nonhuman animals living in the wild could be argued for within animal rights theory (ibid.).

A total hands-off approach can certainly be difficult to implement, as it is challenging to protect natural areas from human intervention. It will be a challenge to protect as many habitats as possible in the sense of a stewardship model, especially if humans cannot generate any direct (large) profit (for themselves) from the establishment of such a protected zone. Hadley (ibid., 17) agrees when he states that "[t]here would be no point in establishing an animal property rights system unless there is to be a change in how animal interests are dealt with when land management decisions are made."

At the moment, humanity is heading towards ever-increasing displacement and decimation of nonhuman animals living in the wild on this planet. If we want to halt this development, habitats must be protected. However, due to the fact that the human population is growing rapidly and that every human being has a right to an adequate standard of living—as codified in Article 26 of the Universal Declaration of Human Rights—interspecies conflicts might also arise regarding habitat concessions. Alastair Gunn (2008, 472) points to a strong human demand: "As Africa's population continues to grow, and habitat shrinks, pressure on wildlife will increase. Africans, like Western environmentalists, are entitled to a materially adequate standard of life. They cannot and should not be expected to protect wildlife if it is against their interest to do so." The establishment and accessibility of reserves, including respectful and (as far as possible) non-disruptive observation of the nonhuman animals living there by visitors, may be an important factor in persuading humans to grant habitat to other species as well. Reserves of this kind certainly have the potential to replace zoos and secure habitats for nonhuman animals, since besides the nonhuman animals living in them, humans could also benefit from the creation of such reserves.

However, interestingly, nonhuman animal property rights may not in fact demand too many concessions of humans in terms of their establishment: property rights of nonhuman animals could be established whilst humans "retain the right to own pets, eat meat, or kill spiders" (Bradshaw 2020, 125). Bradshaw (ibid., 155) elaborates: "A philosophical approach might distinguish different levels of animals based on capacity for pain or intelligence, with primates, but not insects, receiving property rights." That said, Hadley (2015, 105) states that a needs-based animal property rights theory could be extraordinarily ambitious and have far-reaching implications. At the centre of such a needs-based theory would be the claim that all those nonhuman animals who rely on natural resources for the satisfaction of their basic needs have valid claims to property rights, irrespective of species (ibid.). Hadley (ibid., 118) continues: "If the land upon which the animal resides is owned by a human being, then the animal property right must impair the rights of existing landholders in some way." We need to rethink land ownership if we want to think of it in an anti-speciesist way. Hadley (ibid.) states that the "task then becomes one of institutional design: How can we translate the animal property right into the existing property rights system?" and asks himself (ibid., 106) how an "animal property rights framework could be sold to the public." He (ibid.) reflects on the environmental benefits of nonhuman animal property rights: "The environmental benefits [...] can be enhanced by giving rights only to keystone species, such as large carnivores, which have a crucial role to play in maintaining the stability and integrity of ecosystems." Those who argue for interventions to reduce wild animal suffering and who endorse interventions concerning predators[1] might, however, not agree with protecting large carnivores. They would rather give property rights to large herbivores since their presence reduces the amounts of r-strategists,[2] and thus could reduce wild animal suffering (Johannsen 2021).

[1] It is interesting to note that some nonhuman animals are "mesopredators"—they predate and are predated upon (cf. Keulartz 2021, 117).

[2] Broadly speaking, there are two evolutionary reproductive strategies species employ: the K-strategy and the r-strategy (MacArthur and Wilson 1967; Pianka 1970). The K-strategy is used by many mammals, and it ensures species survival by restricting reproduction to a small number of offspring who are intensively cared for. Examples of K-strategists include large mammals such as deer, bears, elephants, etc. The r-strategy, by contrast, is used by many lizards, amphibians, fish, and small mammals, and it ensures species survival through quantity. Instead of restricting reproduction and providing intensive care, r-strategists produce large numbers of uncared-for offspring, the majority of whom die painful deaths shortly after hatching or birth. (Johannsen 2017, 6)

Perhaps animal property rights could be made possible by installing human guardians. Hadley (2015, 16) discusses the idea of installing human guardians "who exercise the powers associated with property rights" with a duty to "act in the best interests of the animals." "Guardians will need practical wisdom, a mature personality and adequate communication skills" (ibid.). A statuary authority would have to be created to oversee this guardian system (ibid., 17). Bradshaw (2020) argues for a similar way of making nonhuman animal property rights a reality, envisioning (ibid., 3) nonhuman animals as "own[ing] land through a trust, managed by a human trustee."

Habitat protection is a shared concern of animal rights activists and environmentalists, because protecting natural habitat for wildlife goes hand in hand with protecting nature. Besides recognizing the property rights of nonhuman animals, one might argue that it then makes sense to install property rights for ecosystems and trees (Hadley 2015, 81): Bradshaw (2020, 8) does indeed argue for such an extension, stating that "the appropriate framing of expanding property interests likely will extend beyond wildlife to ecosystems or nature more broadly." A key philosophical question up for debate is the question of whether intrinsic value is "to be found in individual sentient animals or whole ecosystems" (Hadley 2015, 109). Whilst I tie intrinsic value to interests and sentience in this book, and thus do not believe in an intrinsic value of non-sentient nature, there are humans who have intuitions concerning such an intrinsic value.

Approaches to ecological ethics have already advanced theories of the integrity of natural systems and their claim to unimpeded natural functioning. The idea of a claim by non-living natural entities to their integrity was exemplified, for example, in 2012 by the recognition of a river in New Zealand as a person (mentioned already in Chap. 12), including the declaration of certain claims by the river to the preservation of its natural integrity (cf. Tanasescu 2014, 161). According to such approaches, a river is wronged, for example, if its integrity, i.e. its functioning with regard to its goal (to flow into a sea), is impaired (Schlosberg 2014, 81f; Schlosberg 2007, 149). "Integrity, in this sense, is a state where functioning remains; a violation of integrity undermines function and is the definition of injustice" (Schlosberg 2014, 81).

In some places, "rights of nature" are already being established. In 2006, for example, the Community Environmental Legal Defense Fund (CELDF) in Schuylkill County, Pennsylvania, achieved the inclusion of a right of nature in its legislation (ibid.). This CELDF also inspired the

Ecuadorian government, which agreed in 2008 to legally establish a right of nature to "integral respect for its existence"; more precisely: "the right to 'integral respect for [nature's] life cycles, structures, functions and evolutionary processes'" (ibid.). The establishment of such a right, however, is not primarily about taking into account any "interest" of nature with regard to its integrity, but principally about the interests of humans who benefit from the health and functionality of this natural integrity. In this sense, three "representatives of nature," Alberto Acosta, Esperanza Martínez and Vandana Shiva, sued the oil company BP in 2010 after the oil spill in the Gulf of Mexico, claiming that the oil spill had violated the rights of nature (Tanasescu 2014, 157). The plaintiffs did not expect to win this case; they were more interested in walking little-trodden paths and paving the way for nature's rights to some extent (ibid.).

In 2012, Bolivia also established a protection of the rights of "Mother Nature": "The Bolivian law accords equal rights to nature, which it defines as 'the dynamic living system formed by the indivisible community of all life systems and living beings who are interrelated, interdependent, and complementary, which share a common destiny'" (ibid.). In the course of the World People's Conference on Climate Change and the Rights of Mother Earth, a "Universal Declaration of the Rights of Mother Earth" was drawn up on 22 April 2010 in Cochabamba, Bolivia, which was presented to the UN by Evo Morales. This development may sound "too good to be true"; in fact, it is once again anthropocentric economic engines that are driving this innovation in Bolivia's legislature, since this law to "protect Mother Nature" was used shortly after its establishment to crack down on illegal small-scale mining on the grounds that it pollutes nature (Tanasescu 2014, 159f). When assessing the creation of such a law, one should not overlook the fact that the Bolivian government earns good money from the mining rights it holds; thus, interests other than "Mother Nature's" interest in her own integrity may be at play (ibid.).

The conflict between human interests in an adequate standard of living—and compare here the human right to this enshrined under Article 25 of the Universal Declaration of Human Rights—and the interests of nonhuman animals living in the wild in their habitat, (and thus in their lives), is characterized by a power imbalance. In general, the concept of ownership can prove inspiring and enriching for a debate on habitat protection; the stewardship model addressed by Donaldson and Kymlicka or the guardianship model by Hadley are probably the most promising paths to actually achieve protection of wildlife habitats in practice.

The intuition that it is morally praiseworthy to protect natural habitat for nonhuman animals living in the wild is under siege in the animal suffering discourse that has taken place in animal ethics in the last few years. While there is consensus among activists campaigning for nonhuman animal rights on ending factory farming and transforming food systems, debates persist about our responsibilities towards nonhuman animals living in the wild and whether we could (and should) make meaningful efforts to reduce their suffering. The following section illustrates reflections taking place in the wild animal suffering discourse, and tries to help us understand why we should (perhaps) instead of protecting habitats, go ahead and "burn down the forest."

15.2 But What About Wild Animal Suffering?

> The total amount of suffering per year in the natural world is beyond all decent contemplation. During the minute it takes me to compose this sentence, thousands of animals are being eaten alive; others are running for their lives, whimpering with fear; others are being slowly devoured from within by rasping parasites; thousands of all kinds are dying of starvation, thirst and disease. [...] In a universe of blind physical forces and genetic replication, some people are going to get hurt, other people are going to get lucky, and you won't find any rhyme or reason in it, nor any justice. The universe we observe has precisely the properties we should expect if there is, at bottom, no design, no purpose, no evil and no good, nothing but blind, pitiless indifference. (Dawkins 1996, 131f)

The discussions taking place in the wild animal suffering discourse revolve around the argument that there exists a great deal of natural suffering among nonhuman animals living in the wild and that if we are able to reduce this suffering, we may have a responsibility to take action to do so (e.g. Tomasik 2015). This discourse is heavily influenced by utilitarian ethics, with ethicists stating that if it is the case that in nature, suffering outweighs pleasure (and their lives are "net negative"), it might be better for many wild animals (particularly r-strategists such as insects) not to have come into existence at all (Johannsen 2021; Ng 1995).We should thus, if we can, lower the birth-rates of r-strategists and increase the care invested in each offspring, in order to reduce the number of nonhuman animals whose lives are not worth living (Ng 1995, 271ff). The so-called natural equilibrium in nature, which is a human invention, is understood as "something quite terrible because all species suffer enormously in this situation" (Keulartz 2021, 103). Some argue for the destruction of nature in order to reduce suffering occurring in it. These conclusions run counter to prominent naturalistic values and are thus challenging to convey not only to a lay audience, but also to humans who already advocate for nonhuman animal rights. Many humans might be hesitant when it comes to intervening in nature, not least because "[h]umans have a poor track record of intervening with wildlife" (Bradshaw 2020, 68). We are, of course, currently already intervening heavily in nature, engendering a rapid species extinction due to human activity on this planet, also called a "defaunation process" (Keulartz 2021, 118).

Another aspect in the discourse revolving around wild animal suffering is the "predation problem" (cf. e.g. Dorado 2015; Keulartz 2021). Indubitably, predation causes a considerable amount of suffering on this planet. If we wish to strive towards a world which contains less suffering, should we do something about predation? *Could* we do something about it? Different philosophers provide different answers to this question. According to Tom Regan, we have no duty to "interfere with wildlife," since nonhuman animals living in the wild possess the competence to survive on their own in the wilderness (Keulartz 2021, 100). We have no duty "to assist the sheep against the attack of the wolf, since the wolf neither can nor does violate anyone's rights" (Regan 1983, 285). Regan's argumentation for this statement is that predators are not moral agents, and therefore cannot be held responsible for their actions (Keulartz 2021, 105). Josh Milburn (2015) has provided nuanced reflections on Regan's reasoning. He (Milburn 2015) states that "if the wolf's killing of the deer had taken place in a zoo, then there is some moral agent who is blameworthy in this situation, namely the zookeeper" (Keulartz 2021, 105). While Nussbaum (2010, 512) thinks that it is possibly ethically good to protect prey animals in the wild from predators if we can easily do so (while pointing out that the needs of predators must also be taken into account and the interests involved fairly considered), and that we should, for example, discourage lions from hunting if possible, she (Nussbaum 2010, 537) also states that "only the worst harms to the weak members of a species must be prevented, while other forms of hierarchy can be tolerated." As examples of such "worst harms" Nussbaum mentions "when animal parents attack their young or sick, disabled or old members of a species are treated in an overly harsh manner. In the case of both domestic and 'wild' animals, humans have a duty to intervene to prevent such mistreatment" (ibid.).

Sezgin (2014, 198), however, does not agree with Nussbaum that we should prevent predation if we can. She (ibid.) does not consider it to be appropriate to deny a predator their prey: "The lioness has a right towards us not to snatch the antelope she desperately needs for her cubs to survive. If we wanted to protect the prey from the predator, we would be committing a violation of rights: against the predator, who has a right to live out its life with all the behaviours that are inherent in it." Milburn agrees when he states that "the rights of prey do not generally necessitate intervention" (Keulartz 2021, 105). Contrastingly, Sapontzis (1984, 27f) argues that when it comes to our obligation to prevent harm, it is irrelevant if this harm was caused by a moral agent or not: "Because we routinely hold parents responsible for preventing their 'innocent' young child from tormenting the cat, we also seem to have an obligation to stop the cat from killing birds" (cf. Keulartz 2021, 105).

Utilitarians might argue that "the evil of predation" cannot be eliminated without producing even worse evils and ecological disasters (Keulartz 2021, 100). Sapontzis (1984, 31) argues that we should stop predation "whenever doing so would not occasion as much or more suffering than it would prevent." We already intervene in nature on a large scale, through "agriculture, fishing, industry, building, mining, and of course, through nature conservation" (Keulartz 2021, 102). Tyler Cowen (2003) suggests a few policy measures and "attempts to shift nature's balance of power to the detriment of predators and to the benefit of their victims" (Keulartz 2021, 102). We should, according to Cowen, "stop subsidizing the propagation of carnivores and limit or eliminate programmes to protect endangered carnivores or prevent their extinction. We should make hunting structures against killing carnivores less tight" (Keulartz 2021, 103).

There might be intuitive support for reducing as much suffering as possible in this world, which is what Keulartz (2021) calls "paradise engineering." One central "paradise engineer" of our time is the British philosopher David Pearce, who has founded a transhumanist organization called "Humanity+." In his book *The Hedonistic Imperative*, Pearce (1995) imagines ways in which all suffering of sentient beings could be expunged. He advocates for the limitation or elimination of predation, suggesting two ways to achieve this: extinction and reprogramming (cf. Keulartz 2021, 104). Administering depot-contraception to carnivores could bring about their extinction (Pearce 2009). Donaldson and Kymlicka, who argue for a sovereignty approach when it comes to nonhuman animals living in the wild, would certainly view these suggestions critically, stating that by engaging in paradise engineering, we would turn "nature into a well-managed zoo in which each animal has its own safe enclosure and guaranteed food source" (Donaldson and Kymlicka 2011, 164).

Those who argue for a "hands-off" approach face the criticism that their point of view might be at least in part motivated by speciesism. Imagine (cf. Keulartz 2021, 107 for this comparison) a lioness attacking a human baby. Would we (and should we) intervene to save the baby if we could? Many would intuitively argue that we should. What if the lioness attacks a baby zebra? If we choose not to intervene in this case, is it because we discriminate against beings due to their species membership? Charles Fink (2005, 12, cited in Keulartz 2021, 107) chooses the anti-speciesist path when he states that if sheeps are parts of our moral community, "there is an obligation to protect them from wolves, whether or not wolves violate their rights."

Martha Nussbaum, too, embraces paradise engineering and interventions into nature to reduce violence and suffering between the species to a certain extent, when she discusses the so-called Other Species capability, which is "the capability or entitlement to be able to live with concern for and in relation to animals, plants, and the world of nature" (Keulartz 2021, 109). This capability "calls for the gradual formation of an interdependent world in which all species will enjoy cooperative and mutually supportive relations with one another. Nature is not that way and never has been. So it calls, in a very general way, for the gradual supplanting of the natural by the just" (Nussbaum 2006, 399). Although Nussbaum thinks it's absurd to put all predators in "protective detention" (cf. Keulartz 2021, 110), she—like Peter Singer—contends "that we should protect prey animals from predation if we can do so without such massive, harm-producing interventions" (ibid.).

It is certainly hard (if not impossible) to foresee the effects of interventions when intervening in such complex systems as the ecological systems on this planet (Delon and Purves 2018). Simmons (2009, 26) agrees and thinks that saving nonhuman animals living in the wild "on any large scale would have disastrous ecological consequences." Humans might also be anxious of potentially risking their own welfare, should interventions have effects that endanger the basic resources that humans need in order to live and thrive. Humans, after all, are also biological beings dependent on functioning ecosystems to meet their basic needs.

Some scholars in the wild animal suffering debate argue that since the scope of suffering is so immense, it would be almost impossible to actually worsen the situation for nonhuman animals living in the wild, and we should therefore study how to intervene and then "do something" (Johannsen 2021). Scholars who are concerned about the suffering of nonhuman animals living in the wild call for scientific advancements in the field of welfare biology (cf. Ng 1995), which has a central interest in understanding the wellbeing and suffering of nonhuman animals living in the wild.

An example of how a small-scale, well-intentioned (conservationist) intervention may lead to unforeseen suffering is provided by the following example. In 2007, an attempt was made to release a group of Przewalski's horses into the wild in Mongolia[3]:

> In Mongolia, Christian Oswald, together with the Mongolian government, started the reintroduction in 1992 by importing the first horses. After a long period of acclimatization, they were released in 1997 near Takhin Tal in the Gobi B National Park in southwest Mongolia. Since 1999, the project has been managed by the International Takhi Group. From 2001 onwards, the population developed well and by the end of December 2009, 137 horses were already living in the wild. [...] However, the extremely cold and long winter of 2009/2010 led to a severe setback: around 90 horses died of starvation and cold. It was the harshest winter in 50 years. (Sielmann Stiftung n.d.)

If we interfere in nature, and (for example) abolish predation, we also reduce pressures that form abilities both in the predator and the potential prey: Keulartz (2021, 111) mentions an example: The extinct American cheetah "has played a crucial role in shaping the astounding speed of the pronghorn antelope, among other traits such as visual acuity" (Keulartz 2021, 111). In line with their sovereignty approach, Donaldson and Kymlicka (2011, 166) also bring up this so-called flourishing argument when they state that eliminating risks to be harmed or to suffer "would involve a terrible curtailment of freedom, including the freedom to fully develop and explore one's capabilities. Individual action to protect a human child at the moment of harm contributes to her flourishing; collective action to prohibit the actions or processes that create the risk of harm is likely to undermine human flourishing. So, too, with animals." Similar

[3] Scholars and activists who are eager to reduce wild animal suffering might have welcomed this effort, since large herbivores reduce shrubbery which then again reduces populations of r-strategists (Johannsen 2021).

to Regan, Donaldson and Kymlicka (ibid.) also contend that nonhuman animals living in the wild are "competent to manage their own affairs" (Keulartz 2021, 114). Interventions to protect them would condemn them "to a permanent state of dependency" (ibid.). Although they qualify their statement concerning competency in the sense that members of r-selected species might have less "competent agency" than members of K-selected species (Donaldson and Kymlicka 2011, 177), they insist that "we should still respect the sovereignty of wild animals, including those for whom there is minimal evidence of competent agency" (ibid.).

The important, thought-provoking, and possibly counterintuitive debate about wild animal suffering is currently in its infancy: there is much deliberation, reflection, and empirical research (cf. welfare biology) left to do in the field. After having provided insights into some arguments and debates in this discourse, in the next chapter, I explore the evolving discourse on the inclusion of nonhuman animals in political processes in the multispecies society. Alasdair Cochrane emphasizes the need to elevate discussions on nonhuman animal rights to mainstream political discourse, shifting them from the periphery to the centre. Scholars such as Robert Garner, Sue Donaldson and Will Kymlicka present varying perspectives on political systems that prioritize justice for nonhuman animals, from idealistic visions (Donaldons and Kymlicka) to pragmatic approaches grounded in current societal contexts (Garner). Donaldson and Kymlicka's visions of "Zoopolis" propose citizenship for domesticated nonhuman animals within a multispecies polis, promoting ethical relationships while rejecting extinctionist demands. I propose the ideas of macropolitical representation and micropolitical participation and provide examples of what each of these might look like. I discuss several central terms in this discussion, such as the all-affected principle, and Sue Donaldson's ideas of "microboards" for representing nonhuman animal interest in the wider political process, and "animal agoras" as multispecies political arenas.

References

Bradshaw, Karen. *Wildlife as Property Owners: A new conception of animal rights.* Chicago: University of Chicago Press, 2020.

Castelló, Pablo P. "The Language of Zoodemocracy: Contesting Human Sovereignty Over Animals." PhD thesis. Royal Holloway, University of London, 2022.

Cowen, Tyler. "Policing nature." *Environmental Ethics* 25, no. 2 (2003): 169–182. DOI: https://doi.org/10.5840/enviroethics200325231.

Dawkins, Richard. *River Out of Eden: A Darwinian View of Life*. New York: Basic Books, 1996.

Deckha, Maneesha. *Animals as Legal Beings: Contesting Anthropocentric Legal Orders*. Toronto: University of Toronto Press, 2021.

Delon, Nicolas, and Duncan Purves. "Wild Animal Suffering is Intractable." *Journal of Agricultural and Environmental Ethics* 31, no. 2 (2018): 239–260. DOI: https://doi.org/10.1007/s10806-018-9722-y.

Donaldson, Sue, and Will Kymlicka. *Zoopolis: A Political Theory of Animal Rights*. Oxford: Oxford University Press, 2011.

Donaldson, Sue, and Will Kymlicka. *Zoopolis: Eine politische Theorie der Tierrechte*. Translated by Joachim Schulte. Berlin: Suhrkamp, 2013.

Dorado, Daniel. "Ethical interventions in the wild: An annotated Bibliography." *Relations: Beyond Anthropocentrism* 3, no. 2 (2015): 219–238. DOI: https://doi.org/10.7358/rela-2015-002-dora.

Fink, Charles K. 2005. "The Predation Argument." *Between the Species* 13(5) (2005): 1–15.

Gunn, Alastair S. "Environmental Ethics and Trophy Hunting." In *The Animal Ethics Reader*, edited by Susan J. Armstrong, and Richard G. Botzler, 464–473. New York: Routledge, 2008.

Hadley, John. *Animal Property Rights: A theory of habitat rights for wild animals*. London: Lexington Books, 2015.

Johannsen, Kyle. "Animal Rights and the Problem of r-Strategists." *Ethical Theory and Moral Practice* 20, no. 2 (2017): 333–45. DOI: https://doi.org/10.1007/s10677-016-9774-x.

Johannsen, Kyle. Wild Animal Ethics: The Moral and Political Problem of Wild Animal Suffering. London: Routledge, 2021.

Keulartz, Jozef. "Should the Lion Eat Straw Like the Ox? Animal Ethics and the Predation Problem." In *The International Library of Environmental, Agricultural and Food Ethics Volume 33*: *Animals in Our Midst: The Challenges of Co-existing with Animals in the Anthropocene*, edited by Bernice Bovenkerk, and Jozef Keulartz, 99–122. Cham, Switzerland: Springer, 2021. DOI: https://doi.org/10.1007/978-3-030-63523-7_6.

MacArthur, Robert H., and Edward O. Wilson. *The Theory of Island Biogeography*. Princeton, NJ: Princeton University Press, 1967.

Milburn, Josh. "Rabbits, Stoats and the Predator Problem: Why a strong animal rights position need not call for human intervention to protect prey from predators." *Res Publica* 1, no. 3 (2015): 273–289. DOI: https://doi.org/10.1007/s11158-015-9281-2.

Ng, Yew-Kwang. "Towards welfare biology: Evolutionary economics of animal consciousness and suffering." *Biology and Philosophy* 10, no. 4 (1995): 255–285. DOI: https://doi.org/10.1007/BF00852469.

Nussbaum, Martha C. *Frontiers of Justice: Disability, Nationality, Species Membership*. Cambridge, MA: Harvard University Press, 2006.

Nussbaum, Martha C. *Die Grenzen der Gerechtigkeit: Behinderung, Nationalität und Spezieszuhörigkeit*. Translated by Robin Celikates and Eva Engels. Frankfurt am Main: Suhrkamp, 2010.

Pearce, David. *The Hedonistic Imperative (1995)*. Accessed March 23, 2020. http://www.hedweb.com/hedethic/tabconhi.htm.

Pearce, David. "Reprogramming predators." Published 2009. Last modified 2015. Accessed March 23, 2020. http://www.hedweb.com/abolitionist-project/reprogramming-predators.html.

Pianka, Eric R. "On r- and K-Selection." *American Naturalist* 104 (1970): 592–597 Pleasance, Chris. "Police Dogs to get a Pension Plan." *The Daily Mail*. Last modified November 4, 2013. https://www.dailymail.co.uk/news/article-2487540/Police-dogs-pension-plan-Animals-given-1-500-help-pay-medical-bills-retire-service.html

Rachels, James. "Do Animals Have a Right to Liberty?" In *Animal Rights and Human Obligations*, edited by Tom Regan, and Peter Singer, 205–223. Englewood Cliffs, NJ: Prentice Hall, 1976.

Regan, Tom. The Case for Animal Rights. Berkeley, CA: University of California Press, 1983.

Sapontzis, Steve F. "Predation." *Ethics and Animals* 5, no. 2 (1984): 27–38. DOI: https://doi.org/10.15368/ea.1984v5n2.1.

Schlosberg, David. *Defining Environmental Justice: Theories, Movements and Nature*. Oxford: Oxford University Press, 2007.

Schlosberg, David. "Ecological Justice for the Anthropocene." In *Political Animals and Animal Politics*, edited by Marcel Wissenburg, and David Schlosberg, 75–89. London: Palgrave Macmillan, 2014.

Sezgin, Hilal. *Artgerecht ist nur die Freiheit*. Munich: C.H. Beck, 2014.

Sielmann Stiftung. "Wiederansiedlung von Przewalski-Pferden." Published n.d. Accessed October 11, 2015. sielmann-stiftung.de/ausdruck/projekte/projekte-international/wiederansiedlung-von-przewalski-pferden/.

Simmons, Aaron. "Animals, predators, the right to life and the duty to save lives." *Ethics and the Environment* 14 (2009): 15–27. DOI: https://doi.org/10.2979/ete.2009.14.1.15.

Stone, Christopher D. *Should Trees Have Standing? Law, Morality, and the Environment*. Oxford: Oxford University Press, 2010.

Tanasescu, Mihnea. (2014) "The Rights of Nature: Theory and Practice." In *Political Animals and Animal Politics*, edited by Marcel Wissenburg and David Schlosberg, 150–163. London: Palgrave Macmillan, 2014.

Tomasik, Brian. "The Importance of Wild-Animal Suffering." *Relations Beyond Anthropocentrism* 3, no. 2 (2015): 133–152. DOI: https://doi.org/10.7358/rela-2015-002-toma.

CHAPTER 16

Specific Rights in Nonhuman Animal Rights Declarations (Part 6)

16.1 Article 9: Right to Macropolitical Consideration and Micropolitical Participation

16.1.1 Political Consideration in the Macropolis

Which political theories and systems are most suitable when it comes to including nonhuman animals into our polis? Cochrane (2010, 9), after reviewing the existing literature on the political consideration and inclusion of nonhuman animals, concludes that utilitarian and liberal theories provide the soundest basis. Utilitarianism correctly emphasizes the importance of sentience in deciding who should be considered morally (ibid.). Liberal approaches focus on the individual and their wellbeing and thus, combined with the utilitarian consideration of sentience, can provide an appropriate approach that recognizes the subject as the proper focus of just politics (ibid.). The strengths of both approaches could then be combined in an interest-based approach (ibid.).

Wissenburg (2014, 43) reflects on feudalism as a political form of the human-nonhuman animal relationship. His approach does not correspond to the general tendency of the animal rights movement, which strives for "animal liberation," because feudalism is a system in which exploitation in the form of feudal services is an intrinsic component. Wissenburg's

D. Schneeberger, *Envisioning a Better Future for Nonhuman Animals*, The Palgrave Macmillan Animal Ethics Series,
https://doi.org/10.1007/978-3-031-75332-9_16

contribution is thus not suitable for the formation of a utopian, just, and symbiotic multispecies community. A picture of such a possible society and the norms established there has been painted in rich detail by Sue Donaldson and Will Kymlicka (2011): a picture of a futuristic society that no longer exploits nonhuman animals, but integrates them into a system of respectful cooperation and symbiosis.

As already made clear, Donaldson and Kymlicka (2013, 175) argue for the abolition of the exploitation of nonhuman animals, but view an extinctionist demand for total separation of humans and nonhuman animals critically: "In our view, this [...] demand to end all relationships with domesticated animals is strategically a disaster for the [animal rights] movement. After all, there are many people who have come to care about animal rights precisely because of their relationship with a companion animal." They argue (ibid., 146) that relationships consistent with norms of justice are possible, whereas bringing about the extinction of domesticated animals would be tantamount to us humans abdicating responsibility for them. Donaldson and Kymlicka (ibid., 144) suggest not only that domesticated nonhuman animals should be considered members of a multispecies polis formed by human and nonhuman animals, but even confer citizenship status on these nonhuman animal members: "Giving citizenship status to domesticated animals might ensure that they are treated fairly in a mixed human-animal society."

Donaldson and Kymlicka (2011) separate nonhuman animals into different categories that come with different rights and duties. They divide nonhuman animals in the "zoopolis" they outline into "citizens," "denizens" (or "liminal animals") and "wild animals." Domesticated animals are full citizens, while nonhuman animals living on the fringes of the multispecies society, so-called liminal animals, are "denizens," i.e. semi-citizens. Nonhuman animals who live entirely independently of human civilization form the third category (ibid.). Wissenburg (2014, 36) is of the opinion that the categorization of some nonhuman beings as citizens causes more problems than it is able to solve, listing (ibid., 38) four main criticisms of Donaldson and Kymlicka's extension of the concept of citizen to (some) nonhuman animals. First, as Wissenburg sees it, it undermines the fundamental values of the concept of citizenship; second, it colonizes the private sphere; third, such an extension demands the impossible from nonhuman animals and the improbable from humans; and fourth, Wissenburg dislikes the fact that it seeks to cure the "exploitation of nonhuman animals," which he himself sees as a dubiously constructed designation of the status quo (ibid.).

Instead of conceiving of nonhuman animals living in interspecies communities with humans as citizens, Wissenburg (2014, 14) proposes a middle way, similar to Benz-Schwarzburg (2012, 250): "the either/or dichotomy needs to be replaced by a subtler conception of 'the subject' compatible with a wider range of possible legal statuses for animals." Furthermore, Wissenburg (2014, 14) notes that a convincing argument still has to be made for taking nonhuman animals and their fate out of the private sphere (away from their human "owners" and their private decisions about how to deal with "their" nonhuman animal) and into the public sphere, so that it becomes even more explicitly the state's business, with the state having to control to an even greater degree what happens (or is done) to nonhuman animals living with humans.

It would likely not be entirely reasonable to assume that nonhuman animals would be interested in participating directly in the macropolitical system that humans have built for themselves. This is probably what Nussbaum (2010, 514) means when she says that it there is no harm in nonhuman animals not having a right to vote; it is, however, beyond dispute that they have interests which are (also) negotiated in the macropolitical arena. If nonhuman animal interests were fairly considered in our political discourses, different laws would be passed.

When it comes to political justice in a multispecies polity, Korsgaard's (2011) proposal of a norm inspired by the categorical imperative can again be brought into the discussion. Then, ceteris paribus, only those actions directed at nonhuman animals would be passed that they themselves would agree to. There would then be no reason for animal rights actionism, because the needs and interests of nonhuman animals would already be considered and protected. Although Korsgaard's approach would arguably prohibit actions which would go too far against the interests of other animals, it might also be seen as going too far with regard to political decisions in an interspecies community between humans and nonhuman animals. Just as children—given the choice—might not agree to having to go to school every day, or to having to do homework, it might be ethically good to nevertheless send children to school and support them with homework. Similarly, certain paternalistic decisions can also be ethically good and just vis-à-vis nonhuman animals (cf. Nussbaum 2010, 532).

If there were a nonhuman animal right to political participation, we would have to consider which interests of which nonhuman animals should be taken into account politically, and how. It is clear that we humans cannot and must not persist with the narrow focus on the interests (of individuals) of our species: we are too powerful and have too much moral potential—and thus too much moral responsibility—not to take the interests of nonhuman animals politically into account as well. If we want to liberate ourselves from our excessive speciesism, we must begin to take the interests of living beings of other species more seriously, both politically and ideally, and in the sense of Singer's principle of fair consideration of interests, include them fairly in our decisions.

How can nonhuman animals be considered politically? In what natural contexts are they actually political actors? Aristotle (2003, 1253a2f; my emphasis) states "that the state is a work of nature and man by nature a state-forming creature [*politikon zoon*]." While according to Aristotle, "man" is clearly a "zoon politikon," in the sense of a communal, social, polis-forming animal; he is not, however, the (only) zoon politikon, as one might infer from Aristotle. Politics, in a broader sense, is a cultural form that arises from the nature and form of life of (social) animals and ideally contributes to the stable functioning of a group, including the cohesion of the latter, which is conducive to the flourishing of the individuals as well. To this end, certain political structures and processes must inevitably emerge. Disputes, fights and deliberations about what collective decisions are made are clearly not a fundamentally human phenomenon. Even in the nonhuman fauna there are recognised hierarchies, along with regular changes of power. In many social groups of nonhuman animals which exhibit hierarchical structures, there are political processes and a "rule of law" with laws—albeit unwritten ones—that the members are aware of or are made aware of through "sanctions," which individuals must abide by, if they wish to avoid punishment.

I will leave it to the political scientists to determine whether—as Winston Churchill remarked to the British House of Commons in 1947—"democracy is the worst form of government except for all the others that have been tried" (Foreign, Commonwealth & Development Office 2014). Democracy is arguably the best (i.e. the most just) political system at present, even if it has weaknesses (e.g. its relative inertia in terms of moral and legal progress, for example) and disadvantages (e.g. just because a view or practice is held and endorsed by the majority of a people does not mean that it is not or cannot be ethically deficient (i.e. unjust)). States struggle to establish and maintain democratic structures. In a democracy, the equal value of every human individual and the equal right of participation of every person entitled to vote are recognised. In addition, the welfare and interests of humans belonging to diverse vulnerable groups—for example, (small) children, mentally impaired humans, humans with dementia—are ideally fairly included in political decisions. Individuals who belong to vulnerable groups need special legal protection, as they might be less able (or entirely unable) to advocate for their interests and concerns sufficiently effectively on their own. They are at the mercy of other members of the community who do not belong to these vulnerable groups and their decisions.

In order to protect the vulnerable group of human children internationally, the Convention on the Rights of the Child, ratified in 1990, was adopted. Exactly 18 years later, in 2008, the Convention on the Rights of Persons with Disabilities entered into force. Human children and humans with disabilities constitute two groups whose members are especially vulnerable and dependent on others, which is why there must justly be effective forms of representation of their interests. Laws are thus ideally passed with appropriate consideration for their welfare and flourishing. Every state that has signed the relevant conventions must ensure that the rights of these vulnerable groups are recognised and protected. Children and the severely mentally impaired, who cannot vote, are ideally not ignored; on the contrary, their particular vulnerability is highlighted and adequate protection of their goods and interests is ensured, even if, for various reasons, they are not eligible for direct participation in macropolitical deliberations. Nonhuman animals share those features of vulnerability.

When it comes to justice and fair consideration of interests, it can be assumed that a consideration of all subjective views and opinions of the individuals affected is ethically good; not only in the case of decisions in multispecies communities, but also, for example, when it comes to the fair consideration of opinions, wishes and interests of children, mentally disabled humans, and those with dementia. Out of respect—and above all, for reasons of justice—it is imperative to perceive their opinions and wishes, to try to understand them and to include them in decision-making. Such consideration is required due to the fact that the less powerful individuals of a group are also impacted by the decisions made and actions taken within the group. Opinions expressed by individuals of these groups who (have to) live in dependency and partial heteronomy are to be met with emphatic respect, attention and goodwill. See Article 12 of the universal Convention on the Rights of the Child (my emphasis):

1. States Parties shall ensure to the child who is capable of forming his or her own views *the right to express those views freely in all matters affecting the child* and shall give due weight to the views of the child in accordance with his or her age and maturity.
2. To this end, the child shall, in particular, be given the opportunity to *be heard in any judicial or administrative proceedings affecting the child, either directly or through a representative or appropriate body*, in accordance with national procedural rules.

The universal Convention on the Rights of Persons with Disabilities also emphasizes the importance of freedom of expression for persons with disabilities: "Article 21 of the Universal Convention on the Rights of Persons with Disabilities obliges States Parties to take appropriate measures to *enable persons with disabilities to exercise their right to freedom of expression and opinion* on an equal basis with others *through the forms of communication they choose*" (UN-Behindertenrechtskonvention n.d.; my emphasis).

Concerning how nonhuman animal interests can be taken into account in macropolitical processes, Sue Donaldson (2020) proposes installing "microboards" which consist of humans who translate nonhuman animal interests into politics. Castelló (2022, 207f) points out that humans who are part of these boards must "attune themselves to their animality" and undergo a process of subjective transformation to become "coanimals." In order to become "zoocitizens" in the "zooterrapolis" (cf. also Chap. 7 in this book), most humans would need to undergo such processes (Castelló 2022, 44).

In order to be able to ascribe political rights to nonhuman animals, we need both expertise about those beings and expertise on political theories. Wissenburg (2014, 41) points out that Nussbaum's capability approach is fundamentally suitable for political progress towards nonhuman animal rights, but the bottom line is that a separate list of capabilities must be drawn up for each species. For political scientists and policymakers, a just consideration of nonhuman animal interests means that their craft, which is certainly not an easy one, would become yet more complex (ibid.). Political theories about a good and just society must be replaced or expanded by theories about a multispecies society that is also characterized by anti-speciesism.

If we disregard nonhuman animals and their interests in political decisions and fail to provide adequate protection for them and their interests, they will have to bear the (potentially fatal) suffering that these decisions bring upon them. A more just future society that has incorporated anti-speciesism as a value into its value system has recognised that there are goods and vulnerabilities as well as weal and woe in the nonhuman fauna that must be fairly considered. This applies to the welfare of nonhuman animals living in the wild (Faria 2022), but also to nonhuman animals who (have to) live in the multispecies communities with humans, and to all sentient nonhuman animals who are affected by our actions. The all-affected principle envisions such a respect for all those affected by political decisions. Robert Garner (2016, 117) elaborates on this principle

> which is regarded by many democratic theorists (and green political theorists) as the only possible solution to the boundary problem in political democratic theory. That is, when asked who is to count as a member of the polis, one convincing answer is all of those who are affected by decisions taken by that polis. This principle allows for the inclusion of animals if and when their interests are affected by decisions made. It is clear that many political decisions that are made do impact, often detrimentally and often profoundly, upon animals. Therefore, it follows that animals (more specifically, humans acting for animals) ought to have a say in the making of those decisions.

If we take the all-affected principle seriously, then all the beings affected by our political decisions are members of our multispecies polis. Moreover, also future beings will be affected by those decisions, which means that we should find a way to fairly represent and consider their interests as well. When we as humans do not consider nonhuman animals and their interests and make unjust decisions in our favour that harm them, we are behaving selfishly as a species. Even if those nonhuman animals who live in communities with humans, or even those who do not live with humans but are still affected by their decisions, do not have the right to vote in macropolitical contexts, they and their interests should no longer be unfairly disrespected. In particular, nonhuman animals who live in interspecies communities with humans and who may be most directly affected by their actions and decisions, need rights, guardians and legal representatives so that their lives and property are better protected in the future.[1] Furthermore, they, however, are expressing themselves also politically, namely in the micropolis.

[1] In line with the motto prevalent in critical disability studies "Nothing about us without us," Charlotte Blattner (2018) advises us to make nonhuman animal beings present in our reflections about animal ethics, ideally by their personal presence, but if that is not possible, by surrounding ourselves with pictures of them.

16.1.2 Political Participation in the Micropolis

The smallest possible human-nonhuman animal polity consists of one human and one nonhuman animal. Examples of a typical micropolis between a human and a domesticated nonhuman animal are cat and human, dog and human, or horse and human. Beyond that, there are larger communities that can be groups of different sizes and species. In such communities there may be different forms of government or political systems. Political processes are of great importance in communities, since they are what (co-)creates a community in the first place: "processes of deliberation have 'community-generating power'" (Driessen 2014, 102). Tyranny or dictatorship is also conceivable in interspecies communities, just as more democratic forms are possible. In traditional multispecies communities of this kind (at least in human-dominated public spaces and contexts), it is the human who holds the more powerful position and responsibility for the nonhuman creature. In such cases, the human has a duty of care for the nonhuman animal. It is not only interspecies communities that know democratic processes in daily coexistence. "Political" interspecies communication also exists between humans and nonhuman animals living in the wild, and between groups of diverse animals and species. "Recognising [...] political subjectivities involves seeing deliberations occur at unexpected sites, as part of forms of interactions that are often not formally considered political in nature" (ibid., 97).

In order to foster direct interspecies political deliberation, Sue Donaldson proposes the creation of "animal agoras" which are "spaces of unscripted deliberation where new forms of relating and politically participating can emerge" (Castelló 2022, 207). They are "more than human publics" (Smart 2014) and "new spaces of citizenship [...] where human and animal co-citizens can engage one another in spontaneous, unpredictable encounters, spaces they are empowered to re-shape together" (Donaldson 2020, 713). Eva Meijer (2013, 5) points out that nonhuman animals naturally exercise direct political agency. Cojocaru (2020, 75ff), who proposes a "tiered model of political subjectivity" to better represent nonhuman animals in the human dominated political realm, agrees, stating (ibid., 79) that nonhuman animals act politically, for example, when they flee from the slaughterhouse, resist, or carry out acts of sabotage. Meijer (2013, 32) is convinced that nonhuman animals communicate their wishes, opinions and feelings in their interaction with humans, which should be taken into account in political decisions. As an example of intra- and interspecies language games or linguistic intra- and interspecies rituals, Meijer (ibid., 34) discusses the ritual of greeting each other. Meijer (ibid., 36) also cites the example of Irene Pepperberg and her parrot Alex, who have jointly developed an interspecies lingua franca[2] that contains human words but is neither Irene's nor Alex's mother tongue (ibid.). They thus use a jointly created lingua franca to deliberate in their micropolis. Meijer (ibid., 37) also identifies as an example of political communication the behaviour of chimpanzees at the Welsh Mountain Zoo in Colwyn Bay, Wales who communicate with human visitors through sign language. In the past, these chimpanzees have been observed, for example, asking zoo visitors to open the lock of a door and then a window so that they can escape from their cage. This kind of interspecies communication between the chimpanzees and zoo visitors sometimes works flawlessly. Meijer (ibid.) quotes Peter Dickinson, an employee of the zoo: "I have watched our animals sign to visitors, asking them to carry out certain behaviours. What is more, the visitors react and do exactly what they are told. If a visitor is reproached by a member of staff the excuse is always 'But the chimp asked me to do it!'"

For cases where interspecies communication does not work in an unmediated fashion, Meijer (2013, 42) suggests the use of human or nonhuman animal translators. Meijer (ibid., 43) cites Jane Goodall as an example of such a translator between chimpanzees and humans. Clemens Driessen (2014, 91) notes that in numerous encounters and social contexts, the voice or opinion of nonhuman animals is already included in decision-making in political situations. In order to promote and expand this practice of shared, democracy-like decision-making, which certainly contributes to equity between members of an multispecies micropolis, Driessen (ibid.), like Meijer (2013, 47), advocates that humans should and (for political purposes must) make even more of an effort to understand the expressions and contents of the acts of communication that nonhuman animals engage in. Driessen (2014, 92) speaks of this as being about shared learning and decision-making processes in which "new subjectivities emerge," whose desires and demands must be weighed and whose legitimacy and fulfilment must be decided upon.

As an example of deliberation between humans and nonhumans, Driessen (ibid.) cites the experiences of Jenny Diski (2010), who reports

[2] A lingua franca is "a language used for communication between groups of people who speak different languages" (Cambridge Dictionary n.d.).

that it was her cat's wish, as understood by Diski, that she wanted to be let in and out "like a human," despite the presence of a cat flap. Diski complied with this wish by opening the door for the cat to enable this kind of agency. Several examples of deliberation on an equal footing between humans and nonhuman animals are outlined by Martin Balluch (2014), who also describes situations in which trusting, functioning communication between himself and his dog Kuksi were crucial for successful joint action. Balluch goes on to cite an example of micropolitical decision-making in which Balluch decided how to proceed; because Kuksi trusted him, Balluch was able to save them both from danger. The two were on a hike and found themselves surrounded by a herd of cows who wanted to defend their territory (ibid., 32f); Balluch instructed Kuksi to remain seated in the small group of mountain pines in the middle, surrounded by the cows, while Balluch himself made his way to a protective group of mountain pines at the edge:

> We stood in our tiny mountain pine patch for an hour. Kuksi tried to break out once, ran all over the place between the cow legs—my heart almost stopped—and finally returned unharmed but also without having achieved anything. Then, a plan. I suggested to Kuksi that he should stay behind alone and I would try to make my way about fifty meters to a larger mountain pine field. Then he could follow and we would have a chance to escape […]. No sooner said than done, I stepped to the edge of the mountain pine bush. Glancing back at Kuksi, he looked at me trustingly but remained seated. He had fully understood what my intention was. […] Finally I jumped into the nearest mountain pine bushes of the next field and was saved. Now the hardest task, how could I get Kuksi out of there? […] I waited a good 10 minutes, Kuksi did not move. Then I gave Kuksi the signal to run and jumped out of the bushes with a loud shout […] "Run!, run!" I shouted, so that the cows all looked at me. […] Kuksi was much faster than the cows and so he managed to get into the mountain pine field himself in a large arc unharmed. (Ibid., 33)

In this situation, Kuksi showed great trust in Martin Balluch. This is particularly clear since he remained behind, as Balluch had instructed, in a dangerous situation, even though Balluch had taken flight; Kuksi thus clearly understands acts of communication by Balluch. However, he himself also communicates consciously when it comes to political decisions in their micropolis:

> Kuksi can also give me signs. Once we climbed over a very narrow and steep forest ridge that connects one mountain ridge to the other. When we are travelling together, we take it in turns to go ahead. This time he was walking about twenty meters ahead of me and suddenly came back. He stood cross-wise in front of me, looking at me in that characteristic way he always does when he wants to turn around. I didn't understand why he wanted to turn back. "Are you tired already?" I asked him, but I couldn't imagine that he was. So, I climbed over him and kept walking. Less than twenty meters later, the ridge ended abruptly in a five-metre-high rock step, and the rocks on the left and right also descended so steeply that it was impossible to continue. In fact, we had to turn around. (Ibid., 39)

Balluch (ibid., 60ff) also lists diverse acts of communication that Kuksi uses to ask questions or convey various contents. In Kuksi's and Balluch's micropolis, there are certain social norms that they agreed to adhere to: "[O]ur common social rules are not only guidelines from me, he also took part in working them out. [...] For us, it is not just me alone who decides what happens" (ibid., 69). Kuksi uses various forms of political expression or protest. For example, one of the social rules established between Balluch and Kuksi is that a potentially dirty Kuksi is not allowed to lie down in Balluch's bed without being asked (ibid.) "Kuksi adheres amazingly closely to this rule" (ibid., 69f). "But once when he was angry with me, he went straight to the bed and lay down on it demonstratively. In that case he deliberately broke our rule in protest." Kuksi also tries to influence courses of action through passive resistance or to make his will known: "If he doesn't want to come with me or get into the car, he sometimes engages in passive resistance by going completely limp and no longer moving. He thus communicates to me in a rather reserved way that he does not agree with my decision" (ibid., 62).

Driessen (2014, 99), with reference to Haraway (2007), offers a novel understanding of "responsiveness": When we begin to understand the acts of communication towards us emitted by nonhuman animals as meaningful signals, we become "response-able." Driessen (2014, 99) correctly states that if we do not pay attention to the signals sent by nonhuman animals and do not try to understand them, or use their communicative content only insofar as we need it to control and master them, we are the ones who cannot deliberate sufficiently well, not the nonhuman animals.

In the next chapter, I explore a right to education for nonhuman animals, emphasizing its importance in intra- and interspecies contexts.

Education is not only vital for humans, but also for nonhuman animals. It is naturally provided by parents, peers, and social groups. In a multispecies society, both intraspecies (within species) and interspecies (between species) education are essential. In order to include nonhuman animals as workers in the multispecies society, they also need appropriate vocational education. Human guardians bear the responsibility to ensure that the nonhuman animals they care for receive appropriate education from themselves or qualified teachers, akin to the human right to education, as in Article 26 of the Universal Declaration of Human Rights. Education and learning enable individuals to flourish and find a secure place in the multispecies society.

References

Aristotle. *Politik*. Translated by Olof Gigon. Munich: dtv Verlagsgesellschaft, 2003.

Balluch, Martin. *Der Hund und sein Philosoph: Plädoyer für Autonomie und Tierrechte*. Vienna: Promedia, 2014.

Benz-Schwarzburg, Judith. *Verwandte im Geist – Fremde im Recht – Sozio-kognitive Fähigkeiten bei Tieren und ihre Relevanz für Tierethik und Tierschutz*. Erlangen: Harald Fischer, 2012.

Blattner, Charlotte E. “Animals, the Forgotten Proletariat.” Published December 12, 2018. https://youtu.be/HRINOyJe-90.

Cambridge Dictionary. “Lingua Franca.” Published n.d. Accessed March 27, 2024. https://dictionary.cambridge.org/de/worterbuch/englisch/lingua-franca.

Castelló, Pablo P. “The Language of Zoodemocracy: Contesting Human Sovereignty Over Animals.” PhD thesis. Royal Holloway, University of London, 2022.

Cochrane, Alasdair. *An Introduction to Animals and Political Theory*. New York: Palgrave Macmillan, 2010.

Cojocaru, Maria-Dara. “Sind manche Tiere politischer als andere? Politische Subjekte jenseits liberal-demokratischer Denkgewohnheiten.” *TIERethik* 12, no. 20 (2020): 70–101.

Diski, Jenny. *What I Don’t Know about Animals*. Princeton: Yale University Press, 2010.

Donaldson, Sue. “Animal Agora: Animal Citizens and the Democratic Challenge.” *Theory and Practice* 46, no. 4 (2020): 709–735. DOI: https://doi.org/10.5840/soctheorpract202061296.

Donaldson, Sue, and Will Kymlicka. *Zoopolis: A Political Theory of Animal Rights*. Oxford: Oxford University Press, 2011.

Donaldson, Sue, and Will Kymlicka. *Zoopolis: Eine politische Theorie der Tierrechte*. Translated by Joachim Schulte. Berlin: Suhrkamp, 2013.

Driessen, Clemens. "Animal Deliberation." In *Political Animals and Animal Politics*, edited by Marcel Wissenburg and David Schlosberg, 90–104. London: Palgrave Macmillan, 2014.

Faria, Catia. *Animal Ethics in the Wild: Wild Animal Suffering and Intervention in Nature*. Cambridge: Cambridge University Press, 2022.

Foreign, Commonwealth & Development Office. "The Worst form of Government." Published March 5, 2014. Accessed May 24, 2020. https://blogs.fcdo.gov.uk/petermillett/2014/03/05/the-worst-form-of-government/#:~:text=Winston%20Churchill%20once%20said%20that,winning%20the%20Second%20World%20War.

Garner, Robert. "Animals, Politics and Democracy." In *The Political Turn in Animal Ethics*, edited by Robert Garner, and Siobhan O'Sullivan, 103–119. New York: Rowman & Littlefield, 2016.

Haraway, Donna J. *When Species Meet*. Minneapolis MN: University of Minnesota Press, 2007.

Korsgaard, Christine M. "Interacting with Animals: A Kantian Account." In *The Oxford Handbook of Animal Ethics*, edited by Tom L. Beauchamp, and R. G. Frey, 91–118. Oxford: Oxford University Press, 2011.

Meijer, Eva. "Political Communication with Animals." *Humanimalia* 5, no. 1 (2013): 28–52.

Nussbaum, Martha C. *Die Grenzen der Gerechtigkeit: Behinderung, Nationalität und Spezieszuhörigkeit*. Translated by Robin Celikates and Eva Engels. Frankfurt am Main: Suhrkamp, 2010.

Smith, Christie. "Articulating Ecological Injustices of Recognition." In *Political Animals and Animal Politics*, edited by Marcel Wissenburg and David Schlosberg, 61–74. London: Palgrave Macmillan, 2014.

UN-Behindertenrechtskonvention. "Meinungsfreiheit und Informationszugang." Published n.d. Accessed November 4, 2015. behindertenrechtskonvention.info/meinungsfreiheit-und-informationszugang-3879.

Wissenburg, Marcel, and David Schlosburg. *Political Animals and Animal Politics*. London: Palgrave Macmillan, 2014.

CHAPTER 17

Specific Rights in Nonhuman Animal Rights Declarations (Part 7)

17.1 Article 10: Right to Education

> Neither are we the spokespeople of animals nor are we their saviors [...]: animals are the true teachers and leaders of a movement for interspecies justice. (Blattner et al. 2020a, 17)

Education is not an exclusively human phenomenon, as most nonhuman animals also need a certain amount of education to thrive. Human and nonhuman animal children, by their very nature, receive this oftentimes automatically from their parents and/or other members of their social group. In a multispecies society, we need diverse kinds of education and learning. We can speak of intraspecies learning (humans learn from humans, dogs learn from dogs, etc.) and interspecies learning (humans learn from dogs, and dogs learn from geese, for example). When we as humans open our minds in the multispecies community, we can learn from nonhuman beings and become "other-wise" (O'Doherty 2016, 421).[1]

Two different types of education to which nonhuman animals should have a right are stipulated here. First, there is *intraspecies* education, which

[1] Castelló (2022, 208) exemplifies the transformation we as humans can undergo if we start to become what O'Doherty calls "other-wise." Castelló (ibid.) discusses "the process of subjective transformation that Reingard Spannring experienced with her horse friend Freja"; Spannring herself (2019, 91) illustrates how she joined the horses in their world:

D. Schneeberger, *Envisioning a Better Future for Nonhuman Animals*, The Palgrave Macmillan Animal Ethics Series, https://doi.org/10.1007/978-3-031-75332-9_17

takes place even when the species lives separately from humans, and second, there is *interspecies* education, which is important for the individual to thrive as a member of an interspecies community. In interspecies education, members of different species become teachers and students. At VINE sanctuary, for example, members of the community there teach newcomers the rules of the community (cf. Blattner et al. 2020a). Beings belonging to all kinds of animal species can and will both act as teachers and as students in interspecies education. As humans, we must not underestimate and undervalue the wisdom we can learn from those who belong to other species. We can, moreover, also reflect on what sort of education humans should provide for nonhuman beings. Martha Nussbaum (2010, 534) reflects on this: "Some animals are entitled to proper education. If you don't train a border collie, you neglect it, [...] the same is true for many breeds of horses." She (ibid., 532) is convinced that in some cases, paternalistic decisions can be appropriate and important: "That a horse first gets annoyed about the bridle is no more a problem from this perspective of the capability approach than that children get annoyed about compulsory schooling. Both can be justified by the importance of such education in promoting flourishing and skills in adulthood."

Nonhuman animals living in interspecies contexts with humans must also have a right to a certain level of intraspecies education. For example, every social animal should be able to learn basic intraspecies social skills, so that every animal is fundamentally capable of meeting, befriending and socializing appropriately with conspecifics. Furthermore, growing up and interacting and playing with peers is essential for the individual healthy development of a (social) young individual. Insofar as education is essential for a being to thrive, even in the nonhuman animal kingdom, it can be assumed that the right to education is an entitlement right. An individual has an entitlement to education vis-à-vis their guardians. In intraspecies

I join the little herd as they forage in a large alpine pasture. Slowly moving with them, I observe how they balance their bodies on the steep mountainside by bending the joints of their uphill legs. I can feel my own body doing that, slightly leaning towards the mountainside, relaxing. I hear the rhythmic chewing and smell the scent of the torn grass blades. I get down on all fours to explore the diversity of the grasses myself. I feel the warm, damp ground and find a pillow of wild thyme. I bury my face in it, smell its fragrance and let its tiny twigs touch my skin. I listen to the hooves approaching and suddenly feel Freja's lips exploring my hand. Without looking, I turn my hand to feel her soft nose. In the next moment, a screech draws our attention. The horses and I follow the sound in the sky. Two young buzzards are playing high above us.

contexts, as already mentioned, parents (or other family and group members) by their very nature ideally automatically fulfil the duty of upbringing and education. In interspecies communities, the human guardian must ensure that the individual in their care receives intraspecies and interspecies education. This can be provided, if the guardian is sufficiently qualified to do so, by the guardian themselves, or by suitable teachers (cf. the human right to a minimum level of education in Article 26 of the Universal Declaration of Human Rights).

In the case of dogs and horses, in particular, for example, interspecies education is of great importance for the individual who lives or has to live in an interspecies community. Only if the nonhuman animal does not pose a danger to others (especially also to humans) and if they follow the guidance of their guardian(s) (and thus also adhere to the behavioural norms and rules prevailing there), can the domesticated horse or dog thrive in an interspecies community with humans. When it comes to doubts as to whether nonhuman animals can be socialized into the interspecies community, Donaldson and Kymlicka (2014, 35) advise us to "set aside the dichotomy of unruly animals and reason-governed humans and alarmist fears that admitting animals to the polis will result in unruly beasts taking over public space and trampling everyone who gets in their way." A horse who is to be able and willing to work as a riding or therapy horse needs, for example, to receive extensive vocational education for this "profession" from a capable teacher, because—in the current situation at least—the successful training of a domesticated horse and their ability to work in an interspecies community determines their individual prosperity in this community. As long as equine rights such as the right to life, work, protection in case of occupational disability or sick leave, entitlement to receiving care in case of unemployment and entitlement to (early) pension (if applicable) have not (yet) been implemented, the life of a horse is always particularly under threat if the humans who have trained the horse have not done so successfully.

Nonhuman laboratory animals, who are rehomed and taken in by humans as pets and members of an interspecies community after their use, often have severe problems due to their lack of education. In her 2015 lecture at the Oxford Animal Ethics Summer School, French biologist Laure Hoenen discussed a "fourth R" with regard to animal experimentation—besides the first three, which are: "reduce," "refine," and "replace" (cf. e.g. Neuhaus et al. 2022; Müller 2023). Hoenen offered "rehome" in addition to these, explaining that beagles, for example, who by luck find their way into private households after being used in laboratories are much more difficult and demanding to keep as companion animals than comparable dogs who have received the interspecies education companion dogs need. Human guardians often fail to get along with these former experimental dogs and return them, because the dogs from the laboratory are not house-trained, cannot climb stairs, cannot walk on a leash, are socially impoverished, and are scared of many things due to their previous poor quality of life in the lab.

In the next section, I advocate for recognizing a right to work for nonhuman animals based on their innate need to occupy themselves in a species-appropriate and individually meaningful way. Nonhuman animals exhibit "contrafreeloading" behaviour, actively working to obtain food, for example, even if they had the option to obtain it without working for it, indicating a natural inclination towards engaging in purposeful tasks. Work encompasses intraspecies behaviours such as foraging, hunting, and caring for offspring. In an interspecies context, "animal labour" can be beneficial for both human and nonhuman animals when tasks align with a nonhuman animal's nature and wellbeing. For instance, search dogs find pleasure and meaning in helping humans locate individuals in distress. I discuss the "labour-recognition-transformation thesis" which stands for the hope that the status of a worker could secure better treatment and protection for nonhuman animals. With regard to the ways in which nonhuman animals in the animal industry are used to produce their own bodies and their bodily products, the term "metabolic labour" is discussed critically, since this could entail "humanewashing" exploitative and violent practices. I also touch on "embodied consent" and "assent," addressing how nonhuman animal voices can be taken seriously when it comes to interspecies work.

17.2 Article 11: Right to Work

A right to work can be argued for based on the existence of a strong need for species-appropriate and individually meaningful, nature-appropriate or telos-appropriate activity: "[A]nimals suffer if kept in conditions in which they are without something that they will work hard to obtain" (Dawkins 2006, 36). A phenomenon called "contrafreeloading" has been observed in nonhuman animals living in captivity; this occurs when "animals choose to 'work' for their food (such as manipulating a puzzle feeder) rather than receiving it for free. This could be because of the pleasure found in behavioural occupation" (Browning and Veit 2021, 8). It has been shown that nonhuman animals "such as rats, pigeons and many other species" will "continue to work" or engage in the experimental set-up to solve a task, even if the reward is offered without any participation or performance on their part (Shepherdson 1998, 6). The performance of certain activities can, after all, satisfy certain needs and this satisfaction is generally perceived as pleasant and rewarding. Animals have different types and levels of motivation for certain activities or occupations. Particularly vital (innate) natural behaviours tend to correspond to urgent needs and consequently high motivations to perform these behaviours. Benz-Schwarzburg (2012, 429) points to the possible consequences of frustrating the need to engage in meaningful activities: "If the motivation to perform such behaviours cannot be pursued, stress and frustration can arise. The emergence of behavioural disorders is seen as closely related to animals' failure to pursue their behavioural tendencies and achieve their goals (cf. Shepherdson 1998, 6; Clubb et al. 2006, 12)." As already discussed, "satisfaction of telos results in happiness for an animal" (Rollin 2011, 51). This lends itself to deriving a rationale for a nonhuman animal right to work, provided one accepts the premise that work (that is considered as meaningful by the working individual) is an important component of an animal's telos and that the wellbeing of a being is to be taken into account in decisions affecting it. Sezgin (2014, 154) points out the pleasure that work can bring: "There is a very unspectacular pleasure in the everyday use of our bodies and in doing everyday tasks. And I assume that this is equally present in animals, even more pronounced." Moreover, Nussbaum's list of innate capabilities that nonhuman animals should also have a right to exercise (cf. Nussbaum 2010, 528ff) is also useful when reflecting on a right to work.

Here, I understand "work" to be, firstly, an intraspecies activity involving the use of resources such as energy, strength and/or concentration, which is natural to an animal and is perceived by them as meaningful, which the animal wants to do and does if the animal is not hindered in any way in doing it: "Unlike humans in industrialized societies with a strong division of labour, who are usually more concerned with computers, late buses or annoying callers, most free-living animals spend a lot of time every day in the effort to secure their survival and, to this end, above all their food" (Sezgin 2014, 116). In the sense of Rollin's (2011, 51) "fish gotta swim, birds gotta fly" aphorism, meaningful work is based on the essential natural characteristics of the species or individual. Thus, this concept of intraspecies work includes foraging for food and water; migrating to find food and water sources; various hunting activities (for predators); building nests; tending to, educating and caring for offspring; caring for old, sick and disabled members of the community; protecting and defending the group; guard duty; settling disputes; leading the herd; and making political decisions (i.e. decisions on how to proceed in the group, such as which direction of travel to choose, etc.). Just as in the human world, there is cooperation and division of labour in the nonhuman animal kingdom. There are many activities that are natural for an animal that can be strenuous, and yet are gladly and necessarily carried out by nonhuman animals, in order to provide for themselves or their family and social group. These are activities experienced as rewarding, that are meaningful and correspond to their nature and the tasks they have to fulfil in order to survive and flourish. Blattner (2018) stresses the fact that we should keep in mind that nonhuman animals have a strong interest in engaging in work that is meaningful for themselves, their family, and their social group, which might have nothing to do with the human domain.

In an interspecies community that seeks equitable consideration of all individuals and their interests, work can secondly also be an interspecies activity that benefits humans (and/or theoretically other species). Nonhuman animals work with humans in diverse domains. Police forces in Switzerland, for example, use birds of prey to monitor road accidents, search for missing persons, and intercept drones near airports (Blattner and Bossert 2022). In the northeast of India, Khamtis (a Shan people belonging to the "scheduled tribes") work together with elephants in labour activities such as plouging fields or cutting wood (ibid.).

In the case of interspecies work, work done by nonhuman animals for humans should be in accordance with the nature and inclinations of those doing the work, so as not to (unduly) negatively affect their welfare and the fulfilment of their telos. For example, it can be exciting and rewarding work for search dogs to help humans sniff out contraband or truffles, or find individuals in distress (such as humans buried in avalanches). In any functioning community, symbiotic cooperation is a factor in the cohesion and flourishing of the group and its members. Human-nonhuman animal collaborations are justified under certain circumstances, and can be beneficial for all involved and are—in my view—not fundamentally ethically bad. An example of nonhuman animals living in such symbiotic relationships would be the many domestic cats and dogs who often perform care work and emotional labour in interspecies settings (Blattner et al. 2020c, 8).

Recently, there has been much active discussion in the "animal labour discourse" (cf. e.g. Cochrane 2016; Porcher 2017; Shaw 2018; Blattner et al. 2020b; Taylor and Fraser 2019). Central to this discourse is the "labour-recognition-transformation thesis" which states that "[a]ccess to the social role of 'worker,' and to the rights that go with this status, could play a transformative role in shifting from our current relations of instrumentalization and exploitation to relations of shared membership and cooperation" (Blattner et al. 2020c, 5; cf. also Eisen 2020, 139ff). "Animal labour" could be a "potential site of interspecies world-making, in which humans and animals co-construct meaningful and shared social worlds for themselves" (Kymlicka and Blattner 2022, 34). As mentioned earlier in this book, extending the group of beings who can be "workers" has certain similarities to the "personhood extension project." In both instances, we try to expand certain protections which persons and workers have across the species border, in order to improve the status of nonhuman animals. Both movements are driven by both intuitive and strategic forces.

Pepper (2022, 15) sheds a critical light on nonhuman animal work and contends that some nonhuman animals might "have been better off if they had never had the option [to do this work] in the first place," suggesting (2022, 17) that if one considers "the various dangers and risks (e.g. abuse, violence, overwork, manipulation, coercion, etc.) involved in allowing humans to employ animals, it is not obvious that the benefits to animals are worth the potential costs." Currently, those nonhuman animals who are granted protections which resemble workers' rights is extremely small, or as Pepper (2022, 12) puts it: "[M]ost animal labour in current conditions does not meet the bar of justice." The majority of the nonhuman animals who are currently used by humans could be understood as slaves, as Christopher Sebastian McJetters (2014, 129) points out:

> The very definition of slavery is the treatment of one group as property to be bought, sold, and forced into work by another group. If nonhuman animals are not slaves, are they then free? There are not many animals that I know of that exist in human society who voluntarily engaged in this system. Cows do not clock in and clock out. They don't go home to their families. They don't have conversations in the lunchroom. And the only retirement package available to them at the end of their painful lives is a violent death when their usefulness to us has run out.

The conceptual question of whether nonhuman animals produced for meat and animal products in the animal industry should be considered as workers is an interesting one. Dinesh Wadiwel (2020) thinks that they could be considered as workers who work 24/7 and are continually producing themselves as commodities. Guha-Majumdar (2022, 21) adds that "if 'labour' extends to metabolic processes, there cannot be time outside of labour; intensive confinement 'follows' the animal wherever she goes." This understanding of labour is called "metabolic labour" (Beldo 2017; Porcher 2017; Barua 2017, 2018, 2019). The idea of metabolic labour harbours the danger of the "humanewashing" of violent practices (cf. Niesen 2022, 39). "Some authors, for example, construct the killing of working animals (e.g. sheeps, cows or chickens) as part of a "reciprocal relationship" between humans and animals, a "gift" that the animals give us in return for our care" (Blattner and Bossert 2022, 21). Porcher (2017), for example, argues for the continuation of small-scale animal farming with nonhuman animal "workers" who are then also killed. Blattner and Bossert (2022, 23), however, make clear that the killing of nonhuman animal workers is unacceptable. I myself (Schneeberger 2022) have previously illustrated that nonhuman animals used by humans can be divided into three main categories: products, slaves, and workers, arguing that nonhuman animals in the animal industry are categorized as products and often treated accordingly.

In contrast to products or slaves, workers ideally enjoy certain workers' rights. Working members of democratic social communities have labour, social, sick leave, and pension rights that are justly theirs. Nussbaum (2010, 538) also argues that nonhuman animals performing services in interspecies communities with humans should have a right "to working conditions compatible with dignity and respect." Central workers' rights would, for example, be the right to dissent, the right to prohibition of forced labour, and the right to freely choose one's own work (Blattner et al. 2020c).

Nonhuman animals could consent to working with and for humans using "embodied consent" (Pepper 2022, 13). However, the embodied consent must be called into question when it comes to nonhuman animals who have been trained to "consent" (ibid.). As mentioned in Chap. 11, next to embodied consent, assent can be used if an individual cannot consent needs interests (Blattner 2018). It is already used with humans who have dementia or other mental health conditions. Individuals who cannot consent can, however, often express themselves in other ways (Black et al. 2010).

Scholars in the animal labour discourse such as Kendra Coulter (e.g. Coulter 2017) have discussed what "decent work" or "humane jobs" could look like for nonhuman animals. Other scholars in the animal labour discourse have criticized the "focus on productivity and consumption due to its ecological, social, economic and political damage" (Blattner and Bossert 2022, 28). Blattner and Bossert (ibid.) write about this approach:

"Work for the sake of work must be overcome ("post-work" societies should be established), and instead work should be based on individual interests such as affiliation, creativity, teamwork, etc. and for communal purposes" (cf. also Donaldson and Kymlicka 2020). When it comes to nonhuman animal workers' rights, we should then also include their representation in work councils.

Many humans might have the strong conviction that we are allowed to use nonhuman beings for our ends. Herzberg (1999, 33) is of the opinion that humans are allowed to use nonhuman animals: in his opinion, "human use of milk and eggs, of the skins of dead animals, horseback riding, transporting loads, keeping pets, for example, would be permissible" (Herzberg 1999, 34). He (Herzberg, ibid.) continues: "But because humans also use animals in other ways, burdening, exploiting, torturing or killing them, they are already no longer behaving morally, but culpably." In the sense of Korsgaard's neo-Kantian norm, it can be permissible for nonhuman animals to perform certain tasks (e.g. as companion animals) from which humans benefit, provided that these tasks correspond to the nature of a nonhuman animal: "We may interact with the other animals as long as we do so in ways to which we think it is plausible to think they would consent if they could—that is, in ways that are mutually beneficial and fair, and allow them to live something reasonably like their own kind of life" (Korsgaard 2011, 110). Korsgaard (ibid.) lists types of "animal use" that would fulfil this norm: nonhuman animals could work as companion animals, as help for the disabled and in police operations, and as search, rescue and protection animals. The use of their wool and products such as milk or eggs may also meet the requirements of this norm. Cochrane (2012, 89) agrees that it is ethical to keep nonhuman animals for their physical products, "provided that they have a good quality of life." However, the keeping of nonhuman animals for the extraction of these animal products would have to be strictly limited and the focus of agriculture would have to shift to the production of plant foods (ibid.). Cochrane (ibid.) is also of the opinion that the use of nonhuman animal corpses after their natural death could be ethically permissible. It should be borne in mind that such permission would probably be accompanied by an abuse of them; certain humans would certainly find ways to enable the "natural" death of the nonhuman animals they keep, and thus once again make a business out of killing nonhuman animals. It is, however, questionable as to whether humans living in possible future animal-ethical utopian multispecies communities would want to use or consume animal products anyways.

Which nonhuman animals would be allowed to do which work and which labour rights would have to be granted to them in a fair way would have to be discussed and determined more precisely when drawing up specific animal rights declarations. In the next and final section before the conclusion, I argue for extending entitlement rights such as a right to an adequate standard of living and social benefits to nonhuman animals. Just as we recognize labour rights for humans in welfare states, including limited working hours, fair pay, sickness support, unemployment assistance, paid leave, and pensions, there exists a growing discourse around extending similar rights to nonhuman animals. However, advocating for these positive rights faces challenges, especially given the difficulties in securing even basic human rights consistently across Western states. The speciesist worldview complicates matters, as most humans prioritize human welfare over nonhuman animal interests, reflecting ingrained biases that influence resource allocation.

17.3 Article 12: Right to an Adequate Standard of Living and to Certain Social Benefits (Protection in Case of Unemployment and Disability, Entitlement to Pension)

> In 2013, for example, the Nottingham Police decided to give all police dogs a pension of £500 to cover the costs of food, care, accommodation and medical treatment in recognition of their service. On the occasion of the introduction of the pension scheme for police dogs, Paddy Tipping, Police and Crime Commissioner for Nottingham, said: "We look after the people who work for us who have been police officers and staff, they get a decent retirement and I think it's important the same is done for the dogs. These animals work hard for the police and they are officers in their own right." (Blattner and Bossert 2022, 20f).

The example above shows that as nonhuman animals become respected members of our societies, and our societies are transformed into just multispecies societies, nonhuman animals will not only enjoy recognition as legal subjects before the law, or recognition of their role as workers, but also be granted social benefits that come with being a member of a society in a social state. If we include nonhuman animals in our community, we need to treat these members well and in a just and dignified manner. Just and dignified treatment of the members of a welfare state is recognised as including certain labour rights such as the right to limited working hours, fair pay, support in case of sickness and unemployment, paid leave and pension.

Considering that these entitlement rights are indeed part of the Universal Declaration of Human Rights (under Articles 23–25), but are not consistently guaranteed to a sufficient degree even in Western states, it perhaps becomes clear how difficult the task is for those who wish to advocate for positive nonhuman animal rights, alongside negative basic rights for nonhuman animals. If a being is entitled to social benefits—i.e. benefits provided by the community in which it lives—there is a cost that must be absorbed by the community. It is questionable as to whether humans who are unwilling to grant nonhuman animals negative basic rights such as the right to life at the expense of relatively minor restrictions for themselves are willing to spend (more) money for the welfare of nonhuman animals in the community; that said, fiscally, some things could certainly be done—think of the massive subsidies that currently seep into the animal industry and could be used for other purposes, if the majority of Western humans went vegan, for example.

When nonhuman animals are entitled to certain social benefits that have to be paid for by communal funds, conflicts inevitably arise. Should the available funds be used to support impoverished humans or to save nonhuman animals from being killed for financial reasons? Most humans have firmly entrenched speciesist worldviews: they would probably currently vote to use the financial resources available in a state to help humans (who, from an anti-speciesist point of view, are actually already much better protected legally in the Western world than nonhuman animals), because they share the speciesist intuition that, in general, the life and welfare of a human is worth more than that of a nonhuman animal.

Arguing for a nonhuman animal's right to receive a pension (e.g. in old age, or earlier, due to health reasons) is fraught, especially when the pension system of certain human societies threatens to collapse sooner or later due to demographic changes, or is at least in urgent need of reform. It is thus unclear for some humans themselves how much they will receive in the future in terms of a pension (if anything), or how this will then be financed. In the same way, it is difficult to insist on the nonhuman animal right to basic medical care when humans in the comparably rich USA (for example), despite the advances of Obamacare, still lack affordable, nationwide compulsory health insurance, leaving many of them unable to afford the cost of necessary (and often urgent) medical treatment (DeNavas-Walt et al. 2008). Ideally, all sentient beings—including those who are members of interspecies communities—should, if possible, receive support when in need.

References

Barua, Maan. "Nonhuman Labour, Encounter Value, Spectacular Accumulation: The Geographies of a Lively Commodity." *Transactions* 42 (2017): 274–288. DOI: https://doi.org/10.1111/tran.12170.

Barua, Maan. "Animal Work: Metabolic, Ecological, Affective." *Theorizing the Contemporary: Fieldsights*, (July 26, 2018). https://core.ac.uk/download/pdf/162915794.pdf

Barua, Maan. "Animating Capital: Work, Commodities, Circulation." *Progress in Human Geography* 43, no. 4 (2019): 1–20. DOI: https://doi.org/10.1177/0309132518819.

Beldo, Les. "Metabolic Labor: Broiler Chickens and the Exploitation of Vitality." *Environmental Humanities* 9, no. 1 (2017): 108–128. DOI: https://doi.org/10.1215/22011919-3829154.

Benz-Schwarzburg, Judith. *Verwandte im Geist – Fremde im Recht – Sozio-kognitive Fähigkeiten bei Tieren und ihre Relevanz für Tierethik und Tierschutz*. Erlangen: Harald Fischer, 2012.

Black, Betty S., Peter V. Rabins, Jeremy Sugarman, and Jason H. Karlawish. "Seeking Assent and Respecting Dissent in Dementia Research." *American Journal of Geriatric Psychiatry* 18, no. 1 (2010): 77–85. https://doi.org/10.1097/JGP.0b013e3181bd1de2.

Blattner, Charlotte E. "Animals, the Forgotten Proletariat." Published December 12, 2018. https://youtu.be/HRINOyJe-90.

Blattner, Charlotte E., Sue Donaldson, and Ryan Wilcox. "Animal Agency in Community: A Political Multispecies Ethnography of VINE Sanctuary." *Politics and Animals* 6 (2020a): 1–22.

Blattner, Charlotte E., Kendra Coulter, and Will Kymlicka, eds. *Animal Labour. A New Frontier of Interspecies Justice?* Oxford: Oxford University Press, 2020b.

Blattner, Charlotte E., Kendra Coulter, and Will Kymlicka. "Animal Labour and the Quest for Interspecies Justice." In *Animal Labour. A New Frontier of Interspecies Justice?* edited by Charlotte E. Blattner, Kendra Coulter, and Will Kymlicka, 1–25. Oxford: Oxford University Press, 2020c.

Blattner, Charlotte E., and Leonie N. Bossert. "Tierarbeit – Animal Labour – Der Versuch einer Einordnung." *TIERethik* 24 (2022): 8–38.

Browning, Heather, and Walter Veit. "Freedom and Animal Welfare." *Animals* 11, no. 4 (2021): 1148. DOI: https://doi.org/10.3390/ani11041148.

Castelló, Pablo P. "The Language of Zoodemocracy: Contesting Human Sovereignty Over Animals." PhD thesis. Royal Holloway, University of London, 2022.

Clubb, Ros, Sophie Vickery, and Naomi Latham. "Box 1.1.: Motivation and Motivational Explanations for Stereotypes." In *Stereotypic Animal Behaviour: Fundamentals and Applications to Welfare*, edited by Georgia Mason and Jeffrey Rushen, 12. Cambridge, MA: CABI Publishing, 2006.

Cochrane, Alasdair. *Animal Rights Without Liberation*. New York: Columbia University Press, 2012.

Cochrane, Alasdair. "Labour Rights for Animals." In *The Political Turn in Animal Ethics*, edited by Robert Garner and Siobhan O'Sullivan, 15–32. London: Rowman and Littlefield, 2016.

Coulter, Kendra. "Humane Jobs: A Political Economic Vision for Interspecies Solidarity and Human-Animal Wellbeing." *Politics and Animals* 2, no. 1 (2017): 67–77. DOI: https://doi.org/10.1093/oso/9780198846192.003.0002.

Dawkins, Marian Stamp. "The Scientific Basis for Assessing Suffering in Animals." In *In Defense of Animals: The Second Wave*, edited by Peter Singer, 26–39. Oxford: Blackwell Publishing, 2006.

DeNavas-Walt, Carmen, Bernadette D. Proctor, and Jessica C. Smith. "Income, Poverty, and Health Insurance Coverage in the United States: 2007." *U.S. Census Bureau, Current Population Reports*, 60–235. Washington, DC: U.S. Government Printing Office, 2008.

Donaldson, Sue, and Will Kymlicka. "Unruly Beasts: Animal Citizens and the Threat of Tyranny." *Canadian Journal of Political Science / Revue Canadienne de Science Politique* 47, no. 1 (2014): 23–45.

Donaldson, Sue., and Will Kymlicka. "Animal Labour in a Post-Work Society?" In *Animal Labour: A New Frontier of Interspecies Justice?* edited by Charlotte E. Blattner, Kendra Coulter, and Will Kymlicka, 207–228. Oxford: Oxford University Press, 2020.

Eisen, Jessica. "Down on the Farm: Status, Exploitation, and Agricultural Exceptionalism." In *Animal Labour: A New Frontier of Interspecies Justice?* edited by Charlotte Blattner, Kendra Coulter, and Will Kymlicka, 139–159. Oxford: Oxford University Press, 2020.

Guha-Majumdar, Jishnu. "Slavery, Social Death, and Animal Labor." *Politics and Animals* 8 (2022) 18–26.

Herzberg, Guntolf. "Gerechtigkeit für Tiere." In *Tiere ohne Rechte?* edited by Jan C. Joerden, and Bodo Busch, 29–39. Berlin: Springer, 1999.

Korsgaard, Christine M. "Interacting with Animals: A Kantian Account." In *The Oxford Handbook of Animal Ethics*, edited by Tom L. Beauchamp, and Raymond G. Frey, 91-118. Oxford: Oxford University Press, 2011.

Kymlicka, Will, and Charlotte E. Blattner. "Animal Labour and Animal Poltics: Reply to Critics." *Politics and Animals* 8 (2022): 34–54.

McJetters, Christopher Sebastain. "Slavery. It's still a thing." In *Circles of Compassion: Essays Connecting Issues of Justice*, edited by Will Tuttle, 127–132. Danvers, MA: Vegan Publishers, 2014.

Müller, Nico D. "The 3Rs Alone Will Not Reduce Total Animal Experimentation Numbers: A Fundamental Misunderstanding in Need of Correction." *Journal of Applied Animal Ethics Research* 5, no. 2 (2023): 269–284. DOI: https://doi.org/10.1163/25889567-bja10042.

Neuhaus, Winfried, Birgit Reininger-Gutmann, Beate Rinner, Roberto Plasenzotti, Doris Wilflingseder, Joery De Kock, Tamara Vanhaecke, et al. "The Rise of Three Rs Centres and Platforms in Europe." *Alternatives to Laboratory Animals* 50, no. 2 (2022): 90–120. DOI:https://doi.org/10.1177/02611929221099165.

Niesen, Peter. "Landwirtschaftliche Tierhaltung und die Kategorie der Tierarbeit." *TIERethik* 14, no. 1 (2022): 39–56.

Nussbaum, Martha C. *Die Grenzen der Gerechtigkeit: Behinderung, Nationalität und Spezieszuhörigkeit*. Translated by Robin Celikates and Eva Engels. Frankfurt am Main: Suhrkamp, 2010.

O'Doherty, Damian P. "Feline politics in organization: The nine lives of Olly the cat." *Organization*, 23, no. 3 (2016): 407–33. DOI: https://doi.org/10.1177/1350508416629450.

Pepper, Angie. "Is Animal Labour a Viable Route to Interspecies Justice?" *Politics and Animals* 8 (2022): 12–17. DOI: https://doi.org/10.1111/japp.12436.

Porcher, Jocelyne. *The Ethics of Animal Labor: A Collaborative Utopia*. Cham, Switzerland: Palgrave Macmillan, 2017.

Rollin, Bernard E. *Putting the Horse before Descartes: My Life's Work on Behalf of Animals*. Philadelphia: Temple University Press, 2011.

Schneeberger, Doris. "Nichtmenschliche Tiere in Organisationen – Produkte, Sklav*innen, oder Arbeiter*innen?" *TIERethik* 14, no. 1 (2022): 57–83.

Sezgin, Hilal. *Artgerecht ist nur die Freiheit*. Munich: C.H. Beck, 2014.

Shaw, Rosemary. "A Case for Recognizing the Rights of Animals as Workers." *Journal of Animal Ethics* 8, no. 2 (2018): 182–198. DOI: https://doi.org/10.5406/janimalethics.8.2.0182.

Shepherdson, David J. "Tracing the Path of Environmental Enrichment in Zoos." In *Second Nature: Environmental Enrichment for Captive Animals*, edited by David J. Shepherdson, Jill D. Mellen, and Michael Hutchins, 1–14. Washington: Smithsonian Institution Press, 1998.

Spannring, Reingard. "Mutual Becomings? In Search of an Ethical Pedagogic Space in Human-Horse Relationships: Interdisciplinary Approaches to Curriculum and Pedagogy." In Animals in Environmental Education, edited by Teresa Lloro-Bidart, and Valerie S. Banschbach, 79–94. Cham, Switzerland: Palgrave Macmillan. DOI: https://doi.org/10.1007/978-3-319-98479-7_5.

Taylor, Nik, and Heather Fraser. "The Work of Significant Other/s: Companion Animal Relationships in the Future." In *Companion Animals and Domestic Violence*, edited by Nik Taylor, and Heather Fraser, 185–217. Cham, Switzerland: Palgrave Macmillan, 2019.

Wadiwel, Dinesh J. "The Working Day: Animals, Capitalism, and Surplus Time." In *Animal Labour: A New Frontier of Interspecies Justice?* edited by Charlotte E. Blattner, Kendra Coulter, and Will Kymlicka, 181–206. Oxford: Oxford University Press, 2020.

CHAPTER 18

Conclusion and Outlook

The central research question guiding the reflections in this book were: *Should nonhuman animals have rights, too? And if so, which? And why? And how could those rights be implemented?* In this section, I will try to summarize what my answers would be—at this point in time.[1] In this book, I have argued that nonhuman animals should have rights too, since "what happens to them, matters to them." The argumentation that grounds human animal rights can equally ground nonhuman animal rights. They, too, are born with an inherent dignity, which should be recognised and protected collectively. Which rights should be granted to which nonhuman animals would require an answer to which this book can only provide an approximation. Rights are protective fences around goods, needs, capabilities, and those vary in the animal kingdom. We therefore need to examine each species individually.

Some scholars, such as Donaldson and Kymlicka, argue that it makes a difference whether an animal is or is not a citizen of a multispecies society when it comes to their rights. Intuitively, this differentiation could be useful. When it comes to humans, this is, after all, the case: if I am a citizen of Austria, I might have rights vis-à-vis the state that non-Austrians do not enjoy, for example, the right to vote in Austrian elections. However,

[1] Tom Regan: "[D]raw it with a pencil. Because who knows what you might learn in the coming years" (2014).

D. Schneeberger, *Envisioning a Better Future for Nonhuman Animals*, The Palgrave Macmillan Animal Ethics Series,
https://doi.org/10.1007/978-3-031-75332-9_18

nonhuman animal basic rights similar to human rights would transcend rights that depend on citizenship status. Their aim is to provide protection of the most basic goods independently of citizenship status. Compare, here, Article 2 of the Universal Declaration of Human Rights:

> Everyone is entitled to all the rights and freedoms set forth in this Declaration, without distinction of any kind, such as race, colour, sex, language, religion, political or other opinion, national or social origin, property, birth or other status. Furthermore, no distinction shall be made on the basis of the political, jurisdictional or international status of the country or territory to which a person belongs, whether it be independent, trust, non-self-governing or under any other limitation of sovereignty.

We thus might need universal nonhuman animal rights declarations for each species. That said, additionally, it might make sense to install further rights and duties if we organize in smaller multispecies societies.

The reasoning that inherent dignity of nonhuman animals should be recognised and protected immediately opens up conflicts: Whose dignity? How could, for example, both a predator and a potential prey animal have a right to life (which includes the prima facie right not to be killed)? The predation problem is hard to tackle. We need to invest more time and energy in discussing and debating it. In this book, I argue that all sentient nonhuman animals have the same dignity. This could demand far-reaching protections of all those beings belonging to this group. However, I pair this argumentation with a principle of reasonableness. We would not expect humans to no longer walk on grass, in order to avoid killing ants they might involuntarily step on while doing this. However, I ask humans to follow a plant-based lifestyle, since most humans do not need to kill nonhuman animals to meet their nutritional needs.

In this book, I have provided normative reflections on, empirical descriptions of, and interdisciplinary insights into a right to life; a right to freedom; a right to safety/physical integrity/freedom of torture; a right to recognition as a person/(legal) subject before the law; a right to a human guardian, a legal remedy, and an indirect right of legal action; a right to love, sexuality, procreation, pair-building, family, and friendships; a right to property; a right to political consideration and participation; a right to education; a right to work; a right to an adequate standard of living and certain social benefits. I have pointed out that the reasons why we grant those rights to humans are also applicable to many nonhuman animal species, and we should reflect on how we can ensure that they, too, receive the rights they are actually entitled to.

For the time being, human morality is, overall, not yet sufficiently anti-speciesist to enable such a project. We thus need to strive for societal anti-speciesist change in order to enable the design and implementation of nonhuman animal rights declarations. Technological innovations such as cultivated animal products might accelerate such a development. We, however, also need anti-speciesist education and anti-speciesist institutional reform in various domains. We therefore need humans to make use of their human privilege. Although we all are merely individuals amongst billions of other humans, each of us can still contribute to bringing about the world we envision.

If we reduce our speciesism, our relationship with the nonhuman fauna will become more respectful and less violent. Nonhuman animals will suffer less at the hands of humans and will be able to enjoy better lives. We will become aware of our similarities, and appreciate our different ways of being in the world. As humans, we have abilities which enable us—and I would say even require us—to take the wellbeing and suffering of all into account, and to organize our societies in order to strive for a better world for all. This book cannot illuminate the intricacies of how such a better world might manifest itself in detail. It has, I hope, however provided some illumination of certain aspects. We must continue to engage in collective imagination on how an anti-speciesist society could and should look.

I hope that this book has provided an overview of central academic discourses taking place around nonhuman animal rights. A map providing an overview cannot provide detail, but it can provide us with crucial orientation whenever we enter new terrain. Such a map can therefore be a valuable entry point for those wanting to embark on a new academic journey. Those who already find themselves in the midst of the discourse, and in the detailed discussions, might also at times find it useful to zoom out, and take a look at the bigger picture. Looking at the bigger picture helps us to see what terrain remains underexplored. It is—as I have mentioned before—unfortunately beyond the scope of this book to provide detailed reflections on, for example, whose life should be protected, and in what form, and under which circumstances. There have been nuanced debates of these questions, and we need many more such debates. We will also need people, now and then, to provide an overview of these debates, so that we can take stock of "the bigger picture," and continue our journey upwards and onwards.

This book does not, however, deal with certain aspects which might be central to developing universal nonhuman animal rights declarations. As stressed in this book, we need interdisciplinary collaboration in order to work towards developing such declarations. I have, for example, mentioned religious and theological perspectives on nonhuman animals and animal ethics only in passing. Theologians and scholars of religious animal ethics provide inspiring insights which can fruitfully inform the wider animal ethics discourse. Future research should thus, for example, explore what the secular animal ethics discourse can learn from religious perspectives. I, moreover, also have chosen not to explain in depth the historical development of the Universal Declaration of Human Rights (UDHR). Historians, though, might (for example) address the nature and transformative power of the UDHR and elaborate on what we can learn from its development for our task of formulating nonhuman animal rights declarations. Historians might also discuss how the UDHR has shaped human rights and has influenced the International Covenant on Civil and Political Rights and the International Covenant on Economic, Social and Cultural Rights, both of which came into force in 1976. Experts in the field might then provide valuable insights on the emergence of the Convention on the Rights of Persons with Disabilities; certainly, we can learn valuable lessons from this struggle for the animal cause: for example, how those affected could directly participate in developing declarations to protect them. Legal scholars and legal historians can provide nuanced accounts and in-depth insights as to the legal nature, functions, and role of declarations (as compared to conventions and treaties), as well as on their historical, social, and ethical relevance. The history of slavery and what we can learn from it is another historical topic that is highly relevant when it comes to establishing nonhuman animal rights.

When we look at what we have been discussing so far in the discourse on nonhuman animal rights, we see that there are areas which need to be discussed and developed further. One such area is the wild animal suffering debate, which—although highly important, due to the vast scale of the suffering—struggles to tackle the naturalistic intuitions that many humans (including animal ethicists) have. Another area is the debate on the ethics of keeping nonhuman animals as companions, or the animal labour discourse. These are fascinating discussions which, it is to be hoped, will develop considerably and gather pace in the years to come.

To conclude at this point, I would like to again thank all those who contribute intellectually to these debates and who have engaged in advocacy to bring about interspecies justice. Your work is sorely needed.

There is much work to be done here. (Kymlicka and Blattner 2022, 38)

References

Kymlicka, Will, and Charlotte E. Blattner. "Animal Labour and Animal Poltics: Reply to Critics." *Politics and Animals* 8 (2022): 34–54.

Regan, Tom. Interview on DVD *Live and Let Live*. Film directed by Marc Pierschel. Mindjazz Pictures, 2014.

References

Abbate, Cheryl E. "What it's like, or not like, to Bee." *Between the Species* 26, issue 1 (2023): 1–17.

Acampora, Ralph. "Zoos and Eyes: Contesting Captivity and Seeking Successor Practices." In *The Animal Ethics Reader*, edited by Susan J. Armstrong and Richard G. Botzler, 501–506. New York: Routledge, 2008.

Adams, Carol J. *The Sexual Politics of Meat*. New York: Bloomsbury, 2010.

Adams, Carol J, and Lori Gruen. *Ecofeminism: Feminist Intersections with Other Animals and the Earth*. New York: Bloomsbury, 2014.

Akhtar, A. "The Flaws and Human Harms of Animal Experimentation." *Cambridge Quarterly of Healthcare Ethics* 24, no. 4 (October 2015): 407–419. DOI: https://doi.org/10.1017/S0963180115000079.

Albert Schweitzer Stiftung. "Verbandsklagerecht im Tierschutz." Published n.d.-a. Accessed November 1, 2015. https://albert-schweitzer-stiftung.de/themen/verbandsklagerecht.

Albert Schweitzer Stiftung. "Die Tiere: Enten." Published n.d.-b. Accessed November 7, 2015. albert-schweitzer-stiftung.de/massentierhaltung/enten.

Albrecht, Gary L., Katherine D. Seelman, and Michael Bury, eds., *Handbook of Disability Studies*. Thousand Oaks, CA: Sage, 2000.

American Society for the Prevention of Cruelty to Animals (ASPCA). "New Poll Confirms That Overwhelming Majority of Americans Oppose Horse Slaughter." Last modified February 9, 2022. Accessed February 10, 2023. aspca.org/news/new-poll-confirms-overwhelming-majority-americans-oppose-horse-slaughter.

D. Schneeberger, *Envisioning a Better Future for Nonhuman Animals*, The Palgrave Macmillan Animal Ethics Series,
https://doi.org/10.1007/978-3-031-75332-9

Andrews, Kristin, Gary Comstock, Gillian Crozier, Sue Donaldson, Andrew Fenton, Tyler John, L. Syd M. Johnson, et al. "The Philosopher's Brief on Chimpanzee Personhood." Proposed brief by amici curiae philosophers in support of the petitioner-appellant, Court of Appeals, State of New York, 2018.

Animal Protection Index (API). "About the Animal Protection Index (API)." Published n.d. Accessed August 19, 2021. https://api.worldanimalprotection.org/about.

Animals' Angels Investigations and Advocacy. "The Transport of American Horses to Mexican Slaughter Plants." Accessed August 30, 2015. animalsangels.org/images/stories/pdf/the%20transport%20of%20american%20horses%20to%20mexican%20slaughter%20plants.pdf.

Aristotle. *Aristotle in 23 Volumes: Volume 21: Politics*. Translated by Harris Rackham. Cambridge, MA: Harvard University Press, 1944.

Aristotle. *Politik*. Translated by Olof Gigon. Munich: dtv Verlagsgesellschaft, 2003.

Arkeon. "Arkeon." Published n.d. Accessed March 24, 2024. https://arkeon.bio/.

Arnade, Sigrid. "Sichtbarer den je: Würde und Chancengleichheit – Die Behindertenrechtskonvention und die sexuelle Selbstbestimmung behinderter Mensch." In *Sexualität leben ohne Behinderung: Das Menschenrecht auf sexuelle Selbstbestimmung*, edited by Jens Clausen and Frank Herrath, 35–46. Frankfurt: Kohlhammer, 2013.

Aronowitz, Stanley, Dawn Esposito, William DiFazio, and Margaret Yard. "The Post Work Manifesto." In *Post-Work: The Wages of Cybernation*, edited by Stanley Aronowitz and Jonathan Cutler, 31–80. London: Routledge, 1998.

Ärzte gegen Tierversuche. "Welche und wie viele Tiere leiden in Tierversuchen." Published n.d. Accessed April 10, 2024. aerzte-gegen-tierversuche.de/images/pdf/ausstellunga1.pdf.

Audiovisual Library of International Law. "International Convention for the Regulation of Whaling." Published n.d. Accessed February 10, 2023. https://legal.un.org/avl/ha/icrw/icrw.html.

Bagader, Abubakr Ahmed, Abullatif Tawfik El-Chirazi El-Sabbagh, Mohamad As-Sayyid Al-Glayand, and Mawil Yousuf Izzi-Deen Samarrai. "Environmental Protection in Islam (Part 6 of 7): Conservation of Basic Natural Elements - Plants and Animals." Last modified February 24, 2008. Accessed April 6, 2024. https://www.islamreligion.com/article/pdf/324.

Balcombe, Jonathan. Super Fly: *The Unexpected Lives of the World's Most Successful Insects*. London: Penguin Books, 2021.

Baldwin, Ian T., and Jack C. Schultz. "Rapid Changes in Tree Leaf Chemistry Induced by Damage: Evidence for Communication Between Plants." *Science* 221, no. 4607 (1983): 277–279. DOI: https://doi.org/10.1126/science.221.4607.277.

Balluch, Martin. *Die Kontinuität von Bewusstsein: Das naturwissenschaftliche Argument für Tierrechte*. Vienna: Guthmann-Peterson, 2005.

Balluch, Martin. *Der Hund und sein Philosoph: Plädoyer für Autonomie und Tierrechte*. Vienna: Promedia, 2014.

Balluch, Martin. Presentation given about book *Der Hund und sein Philosoph: Plädoyer für Autonomie und Tierrechte*. University of Salzburg. November 5, 2015.

Barad-Andrade, Judith. "The Dog in the Lifeboat Revisited." *Between the Species* 8, no. 2 (1992): 114–117.

Barron, Andrew B., and Colin Klein. "What insects can tell us about the origins of consciousness." *Proceedings of the National Academy of Sciences of the United States of America* 113, no. 18 (2016a): 4900–4908. DOI:https://doi.org/10.1073/pnas.1520084113.

Barron, Andrew B., and Colin Klein. "Insects have the capacity for subjective experience." *Animal Sentience* 1, no. 9 (2016b): 1–19. DOI: https://doi.org/10.51291/2377-7478.1113.

Barua, Maan. "Lively Commodities and Encounter Value. Environment and Planning D." *Society and Space* 34, no. 4 (2016): 725–744. DOI: https://doi.org/10.1177/0263775815626420.

Barua, Maan. "Nonhuman Labour, Encounter Value, Spectacular Accumulation: The Geographies of a Lively Commodity." *Transactions* 42 (2017): 274–288. DOI: https://doi.org/10.1111/tran.12170.

Barua, Maan. "Animal Work: Metabolic, Ecological, Affective." *Theorizing the Contemporary: Fieldsights*, (July 26, 2018). https://core.ac.uk/download/pdf/162915794.pdf

Barua, Maan. "Animating Capital: Work, Commodities, Circulation." *Progress in Human Geography* 43, no. 4 (2019): 1–20. DOI: https://doi.org/10.1177/0309132518819.

Beauchamp, Tom L., and James F. Childress. *Principles of Biomedical Ethics*. New York: Oxford University Press, 2013.

Bekoff, Marc. *The Emotional Lives of Animals: A Leading Scientist Explores Animal Joy, Sorrow, and Empathy – and Why They Matter*. Novato, CA: New World Library, 2007.

Bekoff, Marc. *Das Gefühlsleben der Tiere*. Translated by Elke Franz. Bernau: animal learn Verlag, 2008.

Bekoff, Marc. *Why Dogs Hump and Bees Get Depressed: The Fascinating Science of Animal Intelligence, Emotions, Friendship, and Conservation*. Novato, CA: New World Library, 2013.

Beldo, Les. "Metabolic Labor: Broiler Chickens and the Exploitation of Vitality." *Environmental Humanities* 9, no. 1 (2017): 108–128. DOI: https://doi.org/10.1215/22011919-3829154.

Bennett, Jane. *Vibrant Matter: A Political Ecology of Things*. Durham, NC: Duke University Press, 2010.

Bentham, Jeremy. *Introduction to the Principles of Morals and Legislation.* Accessed April 26, 2024. https://www.econlib.org/library/Bentham/bnthPML.html.

Benz-Schwarzburg, Judith. *Verwandte im Geist – Fremde im Recht – Sozio-kognitive Fähigkeiten bei Tieren und ihre Relevanz für Tierethik und Tierschutz.* Erlangen: Harald Fischer, 2012.

Benz-Schwarzburg, Judith. "Tierethikerin über tote Giraffe: 'Zoos produzieren überschüssige Tiere.'" Published February 11, 2014. Accessed July 12, 2015. derstandard.at/1389860004472/Tierethikerin-ueber-tote-Giraffe-Zoos-produzieren-ueberschuessige-Tiere.

Bernet Kempers, Eva. "Review: Saskia Stucki, One Rights: Human and Animal Rights in the Anthropocene." *LEOH – Journal of Animal Law, Ethics and One Health* (2023): 24–28. DOI: https://doi.org/10.58590/leoh.2023.002

Binder, Alexandra. "Auch wer sich im Spiegel nicht erkennt, kann Ich-bewusst sein." Vet-Journal (February 2014): 46–49. Accessed October 11, 2015. veterinaerjournal.at/fileadmin/userupload/PDF/V0214/V02_REPORTAGE.pdf

Black, Betty S., Peter V. Rabins, Jeremy Sugarman, and Jason H. Karlawish. "Seeking Assent and Respecting Dissent in Dementia Research." *The American Journal of Geriatric Psychiatry* 18, no. 1 (2010): 77–85. https://doi.org/10.1097/JGP.0b013e3181bd1de2.

Blattner, Charlotte E. "Animals, the Forgotten Proletariat." Published December 12, 2018. https://youtu.be/HRINOyJe-90.

Blattner, Charlotte E., and Leonie N. Bossert. "Tierarbeit – Animal Labour – Der Versuch einer Einordnung." *TIERethik* 24 (2022): 8–38.

Blattner, Charlotte E., Sue Donaldson, and Ryan Wilcox. "Animal Agency in Community: A Political Multispecies Ethnography of VINE Sanctuary." *Politics and Animals* 6 (2020): 1–22.

Blattner, Charlotte E., Kendra Coulter, and Will Kymlicka, eds. *Animal Labour. A New Frontier of Interspecies Justice?* Oxford: Oxford University Press, 2020a.

Blattner, Charlotte E., Kendra Coulter, and Will Kymlicka. "Animal Labour and the Quest for Interspecies Justice." In *Animal Labour. A New Frontier of Interspecies Justice?* edited by Charlotte E. Blattner, Kendra Coulter, and Will Kymlicka, 1–25. Oxford: Oxford University Press, 2020b.

Bliesener, Thomas. *Brockhaus Psychologie – Fühlen, Denken und Verhalten verstehen.* Leipzig: F.A. Brockhaus, 2009.

Boesch, Christophe. *Wild Cultures: A Comparison between Chimpanzee and Human Cultures.* Cambridge: Cambridge University Press, 2012.

Bonnet, Charles. *Traité d'insectologie, ou observations sur les pucerons/Observations sur quelques espèces de vers d'eau douce qui, coupés par morceaux, deviennent autant d'animaux complets.* Paris: Durand, 1745.

Bossert, Leonie. "Feministische und fürsorgeethische Ansätze." In *Handbuch Tierethik. Grundlagen – Kontexte – Perspektiven*, edited by Johann. S. Ach and Dagmar Borchers, 117–122. Heidelberg: J. B. Metzler, 2018.

Bouillon, Barbara. "In-Vitro-Fleisch – Nahrungsmittel der Zukunft." Accessed February 7, 2015. future.arte.tv/de/thema/vitro-fleisch.

Bradshaw, Karen. *Wildlife as Property Owners: A new conception of animal rights.* Chicago: University of Chicago Press, 2020.

Brensing, Karsten. *Persönlichkeitsrechte für Tiere.* Freiburg im Breisgau: Herder, 2013.

British Broadcasting Company (BBC). "French water park Marineland criticised after killer whale dies." Published October 13, 2015. Accessed October 25, 2015. bbc.co.uk/newsbeat/article/34515872/french-water-park-marineland-criticised-after-killer-whale-dies.

Browning, Heather, and Walter Veit. "Freedom and Animal Welfare." *Animals* 11, no. 4 (2021): 1148. DOI: https://doi.org/10.3390/ani11041148.

Bruers, Stijn. "Stijn Bruers – All the Benefits of Veganism (IARC 2015)." Talk given at the International Animal Rights Conference in Esch, Luxembourg, 2015. Published September 12, 2015. youtube.com/watch?v=V6mfX7-Dbbc.

Brunn, Franziska. "Wer wie wir ist, bekommt Rechte?!" Accessed November 1, 2015. tierbefreiung.de/archiv/56/gap_brunn.html.

Bundesgesetzblatt für die Republik Österreich. "291. Verordnung des Bundesministers für Gesundheit über die Kennzeichnung von Schweinen, Schafen, Ziegen, und Equiden." Published September 10, 2009. Accessed November 26, 2015. https://animaldata.com/TKVO.pdf.

Bundesministerium für Ernährung und Landwirtschaft. "Verbot des Kükentotens." Last modified December 4, 2023. Accessed February 20, 2023. bmel.de/DE/themen/tiere/tierschutz/tierwohl-forschung-in-ovo.html.

Bundesministerium für Soziales, Gesundheit, Pflege und Konsumenschutz. "Schwerpunkt: Tiergesundheit und Tierschutz." Published n.d.-a. Accessed October 11, 2015. bmg.gv.at/home/Schwerpunkte/Tiergesundheit/Tierschutz/Ombudsstellen.

Bundesministerium für Soziales, Gesundheit, Pflege und Konsumenschutz. "Häufige Fragen: Chip-Pflicht, Kennzeichnung und Registrierung von Hunde." Published n.d.-b. Accessed November 26, 2015. https://www.sozialministerium.at/home/Service/FAQ_Haeufige_Fragen_/Chip_Pflicht_Kennzeichnung_und_RRegistrierun_von_Hunden.

Burgat, Florence. "The Unexpected Resemblance between Dualism and Continuism, or How to Break a Philosophical Stalemate." In *French Thinking About Animals*, edited by Louisa Mackenzie and Stephanie Posthumus, 49–60. Translated by Dantzel Cenatiempo. East Lansing, MI: Michigan State University Press, 2015.

Burruss, Robert. Cited in "The Experimental Use of Chickens and Other Birds in Biomedical and Agricultural Research: Conclusion – Night of the Living Dead for Birds." Accessed February 4, 2015. upc-online.org/experimentation/experimentalConclusion.htm.

Buschbacher, Anton. "Tierrechte." PhD thesis, University of Salzburg, 2015.

Calarco, Matthew. "Deconstruction is not vegetarianism: Humanism, subjectivity, and animal ethics." *Continental Philosophy Review*, 37, no. 2 (2004): 175–201. DOI: https://doi.org/10.1007/s11007-005-3925-4.

Calarco, Matthew. *Beyond the Anthropological Difference*. Cambridge: Cambridge University Press, 2020.

Cambridge Dictionary. "Lingua Franca." Published n.d. Accessed March 27, 2024. https://dictionary.cambridge.org/de/worterbuch/englisch/lingua-franca.

Cammaerts, Marie-Claire, and Roger Cammaerts. "Are Ants (Hymenoptera, Formicidae) Capable of Self Recognition?" *Journal of Science* 5, no. 7 (2015): 521–532.

Caro, Dario, Steven J. Davis, Simone Bastianoni, and Ken Caldeira. "Global and regional trends in greenhouse gas emissions from livestock." *Climatic Change* 126, no. 1–2 (2015): 203–216. DOI:https://doi.org/10.1007/s10584-014-1197-x.

Castelló, Pablo P. "The Language of Zoodemocracy: Contesting Human Sovereignty Over Animals." PhD thesis. Royal Holloway, University of London, 2022a.

Castelló, Pablo P. "The erasures of Peter Singer's theory, and the ethical need to consider animals as irreducible others." *Philosophy Today*, 66, no. 1 (2022b): 637–653. DOI: https://doi.org/10.5840/philtoday2022325454.

Castelló, Pablo P. "Zoolondopolis." *Animal Studies Journal*, 11(1) (2022c): 121–145. DOI:https://doi.org/10.14453/asj/v11i1.6.

Cavalieri, Paola. "The Animal Debate: A Reexamination." In *In Defence of Animals – The Second Wave*, edited by Peter Singer, 54–68. Oxford: Blackwell Publishing, 2006.

Cavalieri, Paola, and Peter Singer, eds. *Menschenrechte für die Großen Menschenaffen: "Das Great Ape Project."* Munich: Wilhelm Goldmann Verlag, 1993.

Caviola, Lucius, Jim A.C. Everett, and Nadira S. Faber. "The moral standing of animals: Towards a psychology of speciesism." *Journal of Personality and Social Psychology*, 116, no. 6 (2019): 1011–1029. DOI: https://doi.org/10.1037/pspp0000182.

Cetacean Rights. "Declaration of Rights for Cetaceans: Whales and Dolphins." Published May 22, 2010. Accessed November 18, 2015. https://www.cetaceanrights.org.

Claire, Jean K. "Moral Extensionism or Racist Exploitation? The Use of Holocaust and Slavery Analogies in the Animal Liberation Movement." New Political Science 33, no. 3 (2011): 311–333. DOI:https://doi.org/10.1080/07393148.2011.592021.

Clubb, Ros, Sophie Vickery, and Naomi Latham. "Box 1.1.: Motivation and Motivational Explanations for Stereotypes." In *Stereotypic Animal Behaviour:*

Fundamentals and Applications to Welfare, edited by Georgia Mason and Jeffrey Rushen, 12. Cambridge, MA: CABI Publishing, 2006.

Cochrane, Alasdair. "Animal Rights and Animal Experiments: An Interest-Based Approach." *Res Publica* 13 (2007): 293–318. DOI: https://doi.org/10.1007/s11158-007-9037-8.

Cochrane, Alasdair. *An Introduction to Animals and Political Theory*. New York: Palgrave Macmillan, 2010.

Cochrane, Alasdair. *Animal Rights Without Liberation*. New York: Columbia University Press, 2012.

Cochrane, Alasdair. "Cosmozoopolis: The Case Against Group-Differentiated Animal Rights." *Law, Ethics and Philosophy* 1 (2013): 127–141.

Cochrane, Alasdair. "Born in Chains?" In *The Ethics of Captivity*, edited by Lori Gruen, 156–173. Oxford: Oxford University Press, 2014.

Cochrane, Alasdair. "Labour Rights for Animals." In *The Political Turn in Animal Ethics*, edited by Robert Garner and Siobhan O'Sullivan, 15–32. London: Rowman and Littlefield, 2016.

Cojocaru, Maria-Dara. "Sind manche Tiere politischer als andere? Politische Subjekte jenseits liberal-demokratischer Denkgewohnheiten." *TIERethik* 12, no. 20 (2020): 70–101.

Comstock, Gary, G. K. D. Crozier, Andrew Fenton, Tyler John, L. Syd M. Johnson, Robert C. Jones, Nathan Nobis, David M. Peña-Guzmán, et al. "The Philosophers' Brief on Elephant Personhood." New York State Appellate Court, 2020.

Conradi, Elisabeth, and Eva Stollreiter. *Brockhaus Philosophie: Ideen, Denker und Begriffe*. Mannheim: F.A. Brockhaus, 2004.

Convention on the Rights of Persons with Disabilities. Adopted by the United Nations General Assembly in 2006 and entered into force in 2008. Accessed November 1, 2015. behindertenrechtskonvention.at/category/un-brk/deutsch-ubersetzung/.

Convention on the Rights of the Child. Adopted by the United Nations General Assembly in 1989 and entered into force in 1990. Accessed December 29, 2015. unicef.at/fileadmin/media/Kinderrechte/crcger.pdf.

Cooke, Maeve. "Five Arguments for Deliberative Democracy." *Political Studies* 48 (2000): 947–969. DOI: https://doi.org/10.1111/1467-9248.00

Cooke, Steve. "Imagined Utopias: Animals Rights and the Moral Imagination." *Journal of Political Philosophy*, 25 (2017): e1–e18. DOI: https://doi.org/10.1111/jopp.12136.

Cooke, Steve. "The Ethics of Touch and the Importance of Nonhuman Relationships in Animal Agriculture." *Journal of Agricultural and Environmental Ethics* 34, no. 2 (2021): 1–20. DOI:https://doi.org/10.1007/s10806-021-09852-5.

Costello, Kimberly, and Gordon Hodson. "Exploring the roots of dehumanization: The role of animal-human similarity in promoting immigrant humanization." *Group Processes & Intergroup Relations* 13, no. 1 (2010): 3–22. DOI: https://doi.org/10.1177/1368430209347725.

Côté-Boudreau, Frederic. "Inclusive autonomy: A theory of freedom for everyone." PhD thesis, Queen's University, 2016.

Coulter, Kendra. "Humane Jobs: A Political Economic Vision for Interspecies Solidarity and Human-Animal Wellbeing." *Politics and Animals* 2, no. 1 (2017): 67–77. DOI: https://doi.org/10.1093/oso/9780198846192.003.0002.

Cowen, Tyler. "Policing nature." *Environmental Ethics* 25, no. 2 (2003): 169–182. DOI: https://doi.org/10.5840/enviroethics200325231.

Craig, Winston J, and Ann R. Mangels. "Position of the American Dietetic Association: Vegetarian Diets." *Journal of the American Dietetic Association* 109 (2009): 1266–1282. DOI: https://doi.org/10.1016/j.jada.2009.05.027.

Cripps, Elizabeth. "Saving the polar bear, saving the world: Can the capabilities approach do justice to humans, animals and ecosystems?" *Res Publica* 16 (2010): 1–22. DOI: https://doi.org/10.1007/s11158-010-9106-2.

Curley, James P., and Eric B. Keverne. "Genes, brains and mammalian social bonds." *Trends in Ecology and Evolution* 20 (2005): 561–567. DOI: https://doi.org/10.1016/j.tree.2005.05.018.

Curry, Tommy. "Killing Boogeyman: Phallicism and the Misandric Micharacterization of Black Males in Theory." *Res Philosophica* 95, no. 2 (2018): 235–272. DOI: https://doi.org/10.11612/resphil.1612.

Daily News. "Opinion: Macdonalds rattles the hen cage." Published September 12, 2015. Accessed October 20, 2015. nydailynews.com/opinion/peter-singer-mcdonald-rattles-hen-cage-article-1.2357723.

Daily Sabah. "Even after green light, lab-grown meat yet to take off in Singapore." Published March 8, 2023. Accessed March 23, 2024. https://www.dailysabah.com/life/food/even-after-green-light-lab-grown-meat-yet-to-take-off-in-singapore.

Dawkins, Marian Stamp. "The Scientific Basis for Assessing Suffering in Animals." In *In Defense of Animals: The Second Wave*, edited by Peter Singer, 26–39. Oxford: Blackwell Publishing, 2006.

Dawkins, Richard. *River Out of Eden: A Darwinian View of Life*. New York: Basic Books, 1996.

de Boer, A., E. M. van Buel, and G. J. ter Horst. "Love is More Than Just a Kiss: A Neurobiological Perspective on Love and Affection." *Neuroscience* 201 (2012): 114–124.

De Waal, Frans. "The need for bottom-up accounts of chimpanzee cognition." In: *The Mind of the Chimpanzee: Ecological and Experimental Perspectives*, edited by Elizabeth V. Lonsdorf, Stephen R. Ross, and Tetsuro Matsuzawa, 265–276. Chicago: University of Chicago Press, 2008.

De Wit, Mieke, Wouter Kegge, Jochem B. Evers, Marleen H. Vergeer-van Eijk, Paulien Gankema, Laurentius A. C. J. Voesenek, and Ronald Pierika. "Plant neighbor detection through touching leaf tips precedes phytochrome signals." *Proceedings of the National Academy of Sciences of the United States of America* 109.36 (2012): 14705–14710. DOI: https://doi.org/10.1073/pnas.1205437109.

Deckha, Maneesha. "Vulnerability, Equality, and Animals." *Canadian Journal of Women and the Law* 27, no. 1 (2015): 47–70. DOI: https://doi.org/10.3138/cjwl.27.1.47.

Deckha, Maneesha. *Animals as Legal Beings: Contesting Anthropocentric Legal Orders.* Toronto: Toronto University Press, 2020.

Deckha, Maneesha. *Animals as Legal Beings: Contesting Anthropocentric Legal Orders.* Toronto: University of Toronto Press, 2021.

DeGrazia, David. (2006) "On the Question of Personhood beyond Homo Sapiens." In *In Defense of Animals: The Second Wave*, edited by Peter Singer, 40–53. Oxford: Blackwell Publishing, 2006.

DeGrazia, David. "Moral Status As a Matter of Degree?" *The Southern Journal of Philosophy* 46 (2008): 181–198. DOI: https://doi.org/10.1111/j.2041-6962.2008.tb00075.x.

DeGrazia, David. "An interest based model of Moral Status." In *Rethinking Moral Status*, edited by Julian Savulescu, Steve Clarke, and Hazem Zohny, 40–56. Oxford University Press, 2021.

Delon, Nicolas, and Duncan Purves. "Wild Animal Suffering is Intractable." *Journal of Agricultural and Environmental Ethics* 31, no. 2 (2018): 239–260. DOI: https://doi.org/10.1007/s10806-018-9722-y.

DeNavas-Walt, Carmen, Bernadette D. Proctor, and Jessica C. Smith. "Income, Poverty, and Health Insurance Coverage in the United States: 2007." *U.S. Census Bureau, Current Population Reports*, 60–235. Washington, DC: U.S. Government Printing Office, 2008.

Derrida, Jacques. *Of Grammatology.* 4th ed. Translated by Gayatri C. Spivak. Baltimore: John Hopkins University Press, 2016.

Deutscher Tierschutzbund. "Die Haltung von Aquarienfischen." Published n.d. Accessed October 29, 2015. tierschutzbund.de/fileadmin/user_upload/Downloads/BroschuBros/Die_Haltung_von_ Aquarienfischen.pdf.

Diski, Jenny. *What I Don't Know about Animals.* Princeton: Yale University Press, 2010.

Donaldson, Sue. "Animal Agora: Animal Citizens and the Democratic Challenge." *Theory and Practice* 46, no. 4 (2020): 709–735. DOI: https://doi.org/10.5840/soctheorpract202061296.

Dombrowski, Daniel A. "Vegetarianism and the Argument from Marginal Cases in Porphyry." *Journal of the History of Ideas* 45, no. 1 (1984): 141–143. DOI: https://doi.org/10.2307/2709335.

Dombrowski, Daniel A. "Rawls and Animals." *International Journal of Applied Philosophy* 12, no. 1 (1998): 63–77. DOI: https://doi.org/10.5840/ijap19981212.

Donaldson, Sue, and Will Kymlicka. *Zoopolis: A Political Theory of Animal Rights.* Oxford: Oxford University Press, 2011.

Donaldson, Sue, and Will Kymlicka. *Zoopolis: Eine politische Theorie der Tierrechte.* Translated by Joachim Schulte. Berlin: Suhrkamp, 2013.

Donaldson, Sue, and Will Kymlicka. "Unruly Beasts: Animal Citizens and the Threat of Tyranny." *Canadian Journal of Political Science / Revue Canadienne de Science Politique* 47, no. 1 (2014): 23–45.

Donaldson, Sue., and Will Kymlicka. "Animal Labour in a Post-Work Society?" In *Animal Labour: A New Frontier of Interspecies Justice?* edited by Charlotte E. Blattner, Kendra Coulter, and Will Kymlicka, 207–228. Oxford: Oxford University Press, 2020.

Donovan, Josephine. "Feminism and the treatment of animals: From care to dialogue." *Signs: Journal of Women in Culture and Society* 31, no. 2 (2006): 305–329. DOI: https://doi.org/10.1086/491750.

Donovan, Josephine, and Carol J. Adams. *The Feminist Care Tradition in Animal Ethics: A Reader.* New York: Columbia University Press, 2007.

Dorado, Daniel. "Ethical interventions in the wild: An annotated Bibliography." *Relations: Beyond Anthropocentrism* 3, no. 2 (2015): 219–238. DOI: https://doi.org/10.7358/rela-2015-002-dora.

Driessen, Clemens. "Animal Deliberation." In *Political Animals and Animal Politics*, edited by Marcel Wissenburg and David Schlosberg, 90–104. London: Palgrave Macmillan, 2014.

Duve, Karen. *Anständig essen – ein Selbstversuch.* Munich: Goldmann, 2011.

Dyer, Owen. "Prisoners' treatment is "bordering on torture," charity says." *BMJ (Clinical research ed.)* 324, no. 7331 (2002): 187. doi:https://doi.org/10.1136/bmj.324.7331.187

Earp, Brian. D. "Female genital mutilation (FGM) and male circumcision: Should there be a separate ethical discourse?" *Practical Ethics.* Oxford: University of Oxford, 2014. https://www.academia.edu/8817976/Female_genital_mutilation_FGM_and_male_circumcision_Should_there_be_a_separate_ethical_discourse. DOI: https://doi.org/10.13140/2.1.3530.4967.

Earp, Brian. "What is it like to be a bee?" *Think* 16, no. 45 (2017): 43–49. DOI: https://doi.org/10.1017/s1477175616000336.

Eaton, Randall L. "Orcas and Dolphins in Captivity." In *The Animal Ethics Reader*, edited by Susan J. Armstrong and Richard G. Botzler, 497–500. New York: Routledge, 2008.

Eisen, Jessica. "Down on the Farm: Status, Exploitation, and Agricultural Exceptionalism." In *Animal Labour: A New Frontier of Interspecies Justice?* edited by Charlotte Blattner, Kendra Coulter, and Will Kymlicka, 139–159. Oxford: Oxford University Press, 2020.

Eisnitz, Gail. *Slaughterhouse: The Shocking Story of Greed, Neglect, and Inhumane Treatment Inside the U.S. Meat Industry.* Amherst, NY: Prometheus Books, 2006.

euronews.green. "These abandoned apes have a dark story to tell about US testing." Published April 24, 2022. Accessed February 11, 2023. https://www.euronews.com/green/2022/04/21/these-abandoned-apes-have-a-dark-story-to-tell-about-us-animal-testing

European Union (EU). "Richtlinie 2010/63/EU des Europäischen Parlaments und des Rates vom 22. September 2010 zum Schutz der für wissenschaftliche Zwecke verwendeten Tiere." Published October 20, 2010. Accessed October 11, 2015. eur-lex.europa.eu/LexUriServ/LexUriServ.do?uri=OJ:L:2010:276:0033:0079:de:PDF.

Faria, Catia. *Animal Ethics in the Wild: Wild Animal Suffering and Intervention in Nature.* Cambridge: Cambridge University Press, 2022.

Farm Animal Welfare Council. (2009). *Farm Animal Welfare in Great Britain: Past, Present and Future.* London: FAWC, 2009. assets.publishing.service.gov.uk/government/uploads/system/uploads/attachment_data/file/319292/Farm_Animal_Welfare_in_Great_Britain__PastPresent_and_Future.pdf.

Federal Food Safety and Veterinary Office. "Dignity of the animal." Last modified August 11, 2020. Accessed March 26, 2024. https://www.blv.admin.ch/blv/en/home/tiere/tierschutz/wuerde-des-tieres.html#:~:text=The%20Swiss%20Animal%20Welfare%20Act,be%20respected%20when%20handling%20it.

Ferrari, Arianna. "Animal Disenhancement for Animal Welfare: The Apparent Philosophical Conundrums and the Real Exploitation of Animals. A Response to Thompson and Palmer." *Nanoethics* 6 (2012): 65–76. DOI: https://doi.org/10.1007/s11569-012-0139-1.

Fink, Charles K. 2005. "The Predation Argument." *Between the Species* 13(5) (2005): 1–15.

Fomenko, Olena. "Of Mice and Men: PETA's Animal-Friendly Idioms as a Strategy Against Speciesism?" *Acta Linguistica Lithuanica* 88 (2023): 165–180. DOI: https://doi.org/10.35321/all88-08.

Food and Agriculture Organization of the United Nations (FAO), International Fund for Agricultural Development (IFAD), United Nations Children's Fund (UNICEF), World Food Programme (WFP), and World Health Organisation (WHO). *The State of Food Security and Nutrition in the World 2021: Transforming food systems for food security, improved nutrition and affordable healthy diets for all.* Rome: FAO, 2021. DOI: https://doi.org/10.4060/cb4474en.

Foreign, Commonwealth & Development Office. "The Worst form of Government." Published March 5, 2014. Accessed May 24, 2020. https://blogs.fcdo.gov.uk/petermillett/2014/03/05/the-worst-form-of-government/#:~:text=Winston%20Churchill%20once%20said%20that,winning%20the%20Second%20World%20War.

Fothergill, Alastair, and Mark Linfield, directors. *Chimpanzee*. Disneynature. 2012.

Francione, Gary L. "Animal Rights and New Welfarism." In *Encyclopedia of Animal Rights and Animal Welfare*, edited by Mark Bekoff and Carron A. Meaney, 45. Westport, CT: Greenwood Press, 1998.

Francione, Gary L. "Blog: Animal Rights and Domesticated Nonhumans." Last modified January 10, 2007a. Accessed March 28, 2015. abolitionistapproach.com/animal-rights-and-domesticated-nonhumans/.

Francione, Gary L. *Animals, Property and the Law*. Philadelphia: Temple University Press, 2007b.

Francione, Gary L., and Robert Garner. *The Animal Rights Debate: Abolition or Regulation?* New York: Columbia University Press, 2010.

Franklin, Julian H. *Animal Rights and Moral Philosophy*. New York: Columbia University Press, 2005.

Franklin, Sarah, and Donna J. Haraway. "Staying with the Manifesto: An Interview with Donna Haraway." *Theory, Culture & Society* 34, no. 4 (2017): 49–63. DOI: https://doi.org/10.1177/0263276417693.

Frayne, David. *The Refusal of Work*. London: Zed Books, 2015.

Fuhrmann, Dunja. "'Im Grunde genommen werden doch Menschen mit Behinderung im Leben viel mehr behindert, als sie tatsächlich behindert sind!' Ein Interview." In *Sexualität leben ohne Behinderung: Das Menschenrecht auf sexuelle Selbstbestimmung*, edited by Jens Clausen and Frank Herrath, 77–87. Stuttgart: Kohlhammer, 2013.

Galamaga, Adam. *Philosophie der Menschenrechte von Martha C. Nussbaum: Eine Einführung in den Capabilities Approach*. Marburg: Tectum Verlag, 2014.

Gallup, Gordon G. Jr. "Chimpanzees: Self-Recognition." *Science* 167 (1970): 86–87. DOI: https://doi.org/10.1126/science.167.3914.86.

Garner, Robert. *A Theory of Justice for Animals: Animal Rights in a Nonideal World*. Oxford: Oxford University Press, 2013.

Garner, Robert. "Animals, Politics and Democracy." In *The Political Turn in Animal Ethics*, edited by Robert Garner, and Siobhan O'Sullivan, 103–119. New York: Rowman & Littlefield, 2016.

Gerber, Pierre J., Henning Steinfeld, Benjamin Henderson, Anne Mottet, Carolyn Opio, Jeroen Dijkman, Alessandra Falcucci, and Guiseppe Tempio. *Tackling climate change through livestock: A global assessment of emissions and mitigation opportunities*. Rome: Food and Agriculture Organisation of the United Nations (FAO), 2013.

Gillespie, Kathryn. *The Cow with Ear Tag #1389*. Chicago: University of Chicago Press, 2018.

Gillon, R. "Autonomy and the principle of respect for autonomy." *British Medical Journal* (clinical research edition), 290(6484) (1985): 1806–1808. DOI: https://doi.org/10.1136/bmj.290.6484.1806.

Goodall, Jane. "Chimpanzees: Bridging the Gap." In *The Great Ape Project: Equality Beyond Humanity*, edited by Paola Cavalieri and Peter Singer, 10–18. New York: St. Martin's Press, 1993.
Gorz, André. *Reclaiming Work: Beyond the Wage-based Society.* Cambridge: Polity, 1999.
Göthling, Stefan, and Anita Kühnel. "Wer Sexualität erleben möchte, sollte dafür Raum zur Verfügung haben' – Über sexuelle Stellvertretung und sexbezogene Behindertenhilfe." In *Sexualität leben ohne Behinderung: Das Menschenrecht auf sexuelle Selbstbestimmung*, edited by Jens Clausen and Frank Herrath, 102–107. Stuttgart: Kohlhammer, 2013.
Grabolle, Andreas. *Kein Fleisch macht glücklich.* Munich: Goldmann, 2012.
Great Ape Project (GAP). "GAP Relaunch." Published n.d. Accessed October 11, 2015. greatapeproject.de/relaunch.
Greenpeace. "Fischerei-Industrie: Leere Meere – volle Kassen." Published n.d. Accessed November 19, 2015. greenpeace.de/themen/meere/fischerei.
Grossman, Joanna. "Political Lobbying for Animals." In *The Routledge Handbook of Animal Ethics*, edited by Bob Fischer, 516–529. London: Routledge, 2019.
Gruen, Lori. *Entangled Empathy: An alternative ethic for our relationships with animals.* New York, NY: Lantern Books, 2015.
Guardian, The. "'Holocaust on a plate' angers US Jews." Published March 4, 2003. Accessed February 20, 2023. theguardian.com/media/2003/mar/03/advertising.marketingandpr.
Guha-Majumdar, Jishnu. "Slavery, Social Death, and Animal Labor." *Politics and Animals* 8 (2022) 18–26.
Gunn, Alastair S. "Environmental Ethics and Trophy Hunting." In *The Animal Ethics Reader*, edited by Susan J. Armstrong, and Richard G. Botzler, 464–473. New York: Routledge, 2008.
Hadley, John. *Animal Property Rights: A theory of habitat rights for wild animals.* London: Lexington Books, 2015.
Handlin, Linda, Eva Hydbring-Sandberg, Anne Nilsson, Mikael Ejdebäck, Anna Jansson, and Kerstin Unväs-Moberg. "Short-term interaction between dogs and their owners – effects on oxytocin, cortisol, insulin and heart rate: an explanatory study." *Athrozoös* 24 (2011): 301–316. DOI: https://doi.org/10.2752/175303711X13045914865385.
Haraway, Donna J.. "FemaleMan©_Meets_OncoMouse™. Mice into Wormholes: A Technoscience Fugue in Two Parts." In *Modest_Witness@Second_Millennium. FemaleMan©_Meets_OncoMouse™*, 49–118. New York: Routledge, 1997.
Haraway, Donna J. "The Companion Species Manifesto." In *Manifestly Haraway*, 91–198. Minneapolis MN: University of Minnesota Press, 2003.
Haraway, Donna J. *When Species Meet.* Minneapolis MN: University of Minnesota Press, 2007.

Haraway, Donna J. *When Species Meet.* Minneapolis MN: University of Minnesota Press, 2008.

Haraway, Donna J. *Staying with the Trouble: Making Kin in the Chthulucene.* Durham, NC: Duke University Press, 2016.

Heffernan, Claire, Mo D. Salman, and Luke York. "Livestock infectious disease and climate change: A review of selected literature." *CAB Reviews* 7, no. 11 (2013): 1–26. DOI: https://doi.org/10.1079/PAVSNNR20120011

Held, David. *Models of Democracy.* Cambridge: Polity Press, 2006.

Hempolinski, Michael. "Haben wir das Recht, Tiere zu töten? Moralische Aspekte der Tötung von Tieren." In *Tiere ohne Rechte?* edited by Jan C. Joerden, and Bodo Busch, 53–61. Berlin: Springer, 1999.

Hennig, Bettina. *Ich bin dann mal vegan.* Frankfurt am Main: Fischer, 2014.

Herrath, Frank. "Human Right Meets Life Reality: What Disables Sexuality?" In *Sexualität leben ohne Behinderung: Das Menschenrecht auf sexuelle Selbstbestimmung*, edited by Jens Clausen, and Frank Herrath, 19–34. Stuttgart: Kohlhammer, 2013.

Herzberg, Guntolf. "Gerechtigkeit für Tiere." In *Tiere ohne Rechte?* edited by Jan C. Joerden, and Bodo Busch, 29–39. Berlin: Springer, 1999.

Hoerster, Norbert. *Haben Tiere eine Würde?* Grundfragen der Tierethik. Munich: Beck, 2004.

Horta, Oscar. "What is Speciesism?" *Journal of Agricultural and Environmental Ethics* 23 (2010): 243–266. DOI: https://doi.org/10.1007/s10806-009-9205-2.

Horta, Oscar. "The Scope of the Argument from Species Overlap." *Journal of Applied Philosophy* 31, no. 2 (2014): 142–54. DOI: https://doi.org/10.1111/japp.12051.

Hoy, Steffen, ed. *Nutztierethologie.* Stuttgart: Verlag Eugen Ulmer, 2009.

Hühner Haltung. "Federpicken." Published n.d.-a. Accessed September 8, 2015. huehner-haltung.de/gesundheit/federpicken.html.

Hühner Haltung. "Das produktive Huhn." Published n.d.-b. Accessed September 27, 2015. huehner-haltung.de/haltung/produktive-huhn.html

Humane Society of the United States, The. "Warning: Anti-Fur Brochure." Last modified 2000. https://web.archive.org/web/20081203133831/http://files.hsus.org/web-files/PDF/AntiFurWarningBro_2000.pdf.

Humane Society of the United States, The. "New York Center Research Chimps Abandoned in Liberia." Published September 2015. Accessed November 4, 2015. humanesociety.org/news/magazines/2015/09-10/new-york-blood-center-research-chimps-abandoned-in-liberia.html.

Hurlemann, René. Cited November 27, 2013. In "Forscher der Ruhr-Uni Bochum entdecken Hormon, das Männer treu macht." Accessed Febraury 21, 2015. derwesten.de/staedte/bochum/forscher-der-ruhr-uni-bochum-entdecken-hormon-das-maenner-treu-macht-id8709690.

Hursthouse, Rosalind. *Ethics, Humans and Other Animals: An Introduction with Readings*. London: Routledge, 2000.

Hursthouse, Rosalind. "Virtue ethics and the treatment of animals." In *The Oxford Handbook of Animal Ethics*, edited by Tom L. Beauchamp, and Raymond G. Frey, 119–143. Oxford: Oxford University Press, 2011.

Insel, Thomas R., and Lawrence E. Shapiro. "Oxytocin receptor distribution reflects social organization in monogamous and polygamous voles." *Proceedings of the National Academy of Sciences of the United States of America* 89, no. 13 (1992): 5981–5985. DOI: https://doi.org/10.1073/pnas.89.13.5981.

Insel, Thomas R., Zuoxin Wang, and Craig Ferris. "Patterns of brain vasopressin receptor distribution associated with social organisation in microtine rodents." *The Journal of Neuroscience* 14, no. 9 (1994): 5381–5392. DOI: https://doi.org/10.1523/JNEUROSCI.14-09-05381.1994.

International Labor Office. "Global Estimates of Modern Slavery." Published 2017. Accessed March 23, 2024. https://www.ilo.org/wcmsp5/groups/public/@dgreports/@dcomm/documents/publication/wcms_575540.pdf.

Irvine, Leslie. *If You Tame Me: Understanding our connection with animals*. Philadelphia, PA: Temple University Press, 2004.

Ja! Natürlich. "Natürlich Bio-Eier vom Mooddorfer Haushuhn." Published n.d. Accessed September 24, 2015. janatuerlich.at/JaNatuerlich/Bio_Eier_vom_Moosdorfer_Haushuhn/Bio_Eier_vom_ Moosdorfer_Haushuhn/Content.aspx.

Jaeckle, Peter. *Amazing Pigs – Mysterious Boar*. Culver City, CA: United Seabears Corporation, 2011.

Johannsen, Kyle. "Animal Rights and the Problem of r-Strategists." *Ethical Theory and Moral Practice* 20, no. 2 (2017): 333–45. DOI: https://doi.org/10.1007/s10677-016-9774-x.

Johannsen, Kyle. Wild Animal Ethics: The Moral and Political Problem of Wild Animal Suffering. London: Routledge, 2021.

Julius, Henri, Andrea Beetz, Kurt Kortschal, Dennis C. Turner, and Kerstin Unväs-Moberg. *Bonding with animals: Psychological and neurobiological foundations of animal-assisted interventions*. Göttingen: Hogrefe Verlag, 2014.

Kalampokas, Emmanouil, Theodoros Kalampokas, Chrisostomos Sofoudis, Efthymios Deligeoroglou, and Dimitrios Botsis. "Diagnosing arthrogryposis multiplex congenita: a review." *ISRN Obstetrics and Gynecology* 2012, no. 3 264918 (September 2012) DOI:https://doi.org/10.5402/2012/264918.

Kallhoff, Angela. *Ethischer Naturalismus nach Aristoteles*. Paderborn: Mentis, 2010.

Kemmerer, Lisa. "Investigating intersections: Exploring the growth and expansion of anymal liberation." *Green Theory and Praxis* 10, no. 3 (2016): 53–76.

Keulartz, Jozef. "Should the Lion Eat Straw Like the Ox? Animal Ethics and the Predation Problem." In *The International Library of Environmental, Agricultural and Food Ethics Volume 33: Animals in Our Midst: The Challenges of Co-existing with Animals in the Anthropocene*, edited by Bernice Bovenkerk, and Jozef Keulartz, 99–122. Cham, Switzerland: Springer, 2021. DOI: https://doi.org/10.1007/978-3-030-63523-7_6.

Khait, Itzhak, Ohad Lewin-Epstein, Raz Sharon, Kfir Saban, Revital Goldstein, Yehuda Anikster, Yarden Zeron, *et al.* "Sounds emitted by plants under stress are airborne and informative." *Cell* 186, no. 7 (March 2023): 1328–1336. DOI: https://doi.org/10.1016/j.cell.2023.03.009.

Kheel, Marti. "From Heroic to Holistic Ethics: The Ecofeminist Challenge." In *Ecofeminism: Women, Animals, Nature*, edited by Greta Gaard, 243–271. Philadelphia, PA: Temple University Press, 1993.

Kim, Claire Jean. 2011. "Moral Extensionism or Racist Exploitation? The Use of Holocaust and Slavery Analogies in the Animal Liberation Movement." *New Political Science* 33 (3): 311–33. doi:https://doi.org/10.1080/07393148.2011.592021.

Knoepffler, Nikolaus. *Angewandte Ethik.* Cologne: Böhlau, 2010.

Knutsson, Simon. "Reducing suffering among invertebrates such as insects: Policy paper." *Sentience Politics* 1 (2016): 1–18.

Ko, Aph, and Syl Ko. *Aphro-ism: Essays on Pop Culture, Feminism, and Black Veganism from Two Sisters.* Brooklyn, NY: Lantern Books, 2017.

Ko, Aph. *Racism as Zoological Witchcraft: A Guide to Getting Out.* New York: Lantern Books, 2019.

Ko, Syl. "'By 'Human,' Everybody Just Means 'White.'" In *Aphro-ism: Essays on Pop Culture, Feminism, and Black Veganism from Two Sisters*, edited by Aph Ko, and Syl Ko, 20–27. New York: Lantern Books, 2017.

Kompatscher, Gabriele. Lecture entitled "Human-Animal Studies: New Perspectives of Animal-Human Relationships." University of Salzburg. December 8, 2015.

Kohda, Masanori, Redouan Bshary, Naoki Kubu, Satoshi Awata, Will Sowersby, Kento Kawasaka, Taiga Kobayashi, and Shumpei Sogawa. "Cleaner fish recognize self in a mirror via self-face recognition like humans." *Proceedings of the National Academy of Sciences of the United States of America*, 120, no. 7 (February 2023) e2208420120. DOI: https://doi.org/10.1073/pnas.2208420120.

Köllen, Thomas, and Doris Schneeberger. "Avoiding unnecessary suffering: Towards a moral minimum standard for humans' responsibility for animal welfare." *Business Ethics, Environment and Responsibility* 32, no. 4 (2023): 1139–1149. DOI: https://doi.org/10.1111/beer.12565.

Kornum, Birgitte R., and Gitte M. Knudsen. "Cognitive testing of pigs (*Sus scrofa*) in translational biobehavioural research." *Neuroscientific Biobehavioural Review.* 35, no. 3 (2010): 437–51. DOI: https://doi.org/10.1016/j.neubiorev.2010.05.004.

Korsgaard, Christine M. "Interacting with Animals: A Kantian Account." In *The Oxford Handbook of Animal Ethics*, edited by Tom L. Beauchamp, and Raymond G. Frey, 91–118. Oxford: Oxford University Press, 2011.

Kotrschal, Kurt. "Warum Menschen mit anderen Tieren sozial sein wollen und können." In *Tiere: Der Mensch und seine Natur*, edited by Konrad Paul Liessmann, 333–359. Vienna: Paul Zsolnay, 2013.

Kymlicka, Will, and Sue Donaldson. "Animal Rights, Multiculturalism, and the Left." *Journal of Social Philosophy* 45, no. 1 (2014): 116–135. DOI: https://doi.org/10.1111/josp.12047.

Kymlicka, Will. "Membership Rights for Animals." *Royal Institute of Philosophy Supplement* 91 (2022): 213–244. DOI: https://doi.org/10.1017/S1358246122000078.

Kymlicka, Will, and Charlotte E. Blattner. "Animal Labour and Animal Poltics: Reply to Critics." *Politics and Animals* 8 (2022): 34–54.

Langley, Gill. (2009) "The validity of animal experiments in medical research." Last modified 2009. Last accessed 30 April 2024. https://www.wellbeingintlstudiesrepository.org/cgi/viewcontent.cgi?article=1050&context=acwp_arte#:~:text=by%20Gill%20Langley&text=So%2C%20in%20medical%20research%2C%20experiments,lesser%20depending%20on%20the%20model, 1–6.

Limon, Georgina, Javier Guitian, and Neville G. Gregory. "A review of the humaneness of puntilla as a slaughter method." *Animal Welfare* 21, no. S2 (2012): 3–8. DOI: https://doi.org/10.7120/096272812X13353700593248.

Lloro-Bidart, Teresa. "A Feminist Posthumanist Multispecies Ethnography for Educational Studies." *Educational Studies* 54, no. 3 (2018): 253–70. DOI: https://doi.org/10.1080/00131946.2017.1413370.

Locke, John. *Versuch über den menschlichen Verstand*. Band I: Buch I und II. Translated by Carl Winckler. Hamburg: Meiner, 2000.

Lovejoy, Arthur Oncken. *The Great Chain of Being: A study of the history of an idea*. Cambridge, MA: Harvard University Press, 1936.

MacArthur, Robert H., and Edward O. Wilson. *The Theory of Island Biogeography*. Princeton, NJ: Princeton University Press, 1967.

MacClellan, Joel. "Size Matters: Animal Size, Contributory Causation, and Ethical Vegetarianism." *Journal of Animal Ethics*, 3 no. 1 (2013): 57–68. DOI: https://doi.org/10.5406/janimalethics.3.1.0057.

MacKinnon, Catharine A. "Rape Redefined." *Harvard Law & Policy Review*, 10, no. 2 (2016): 431–478. DOI: https://doi.org/10.4159/9780674977761-026.

MacKinnon, Catharine A. "Of Mice and Men: A Fragment on Animal Rights." In *The Feminist Care Tradition in Animal Ethics*, edited by Josephine Donovan and Carol J. Adams, 316–332. New York: Columbia University Press, 2007.

Maeckelbergh, Marianne. "Doing is Believing: Prefiguration as Strategic Practice in the Alterglobalization Movement." *Social Movement Studies* 10, no. 1 (2011): 1–20. DOI: https://doi.org/10.1080/14742837.2011.545223.

Mann, Thomas. *The Magic Mountain*. Translated by Helen Tracy Lowe-Porter. New York: Vintage, 1996.

Marino, Lori, and Christina M. Colvin. "Thinking Pigs: A Comparative Review of Cognition, Emotion, and Personality in *Sus domesticus*." *International Journal of Comparative Psychology* 28 Article 23589 (2015). DOI: https://doi.org/10.46867/ijcp.2015.28.00.04

Markell, Patchen. *Bound by Recognition*. Princeton, NJ: Princeton University Publishing, 2003.

Markell, Patchen. "Recognition and Redistribution." In *The Oxford Handbook of Political Theory*, edited by John S. Dryzek, Bonnie Honig, and Anne Phillips. Oxford: Oxford University Press, 2008.

Maslow, Abraham. "A Theory of Human Motivation." *Psychological Review* 50, no. 4 (1943): 370–391. DOI: https://doi.org/10.1037/h0054346.

Matsuzawa, Tetsuro. "Q&A." *Current Biology* 19, no. 8 (2009) 310–312. DOI: https://doi.org/10.1016/j.cub.2009.02.001.

McBurney, Scott. "Congenital Limb Deformity in a Red Fox." *Canadian Cooperative Wildlife Health Centre* 6, no. 1 (1999): 9–10.

McCartney, Paul. "'If slaughterhouses had glass walls, everyone would be vegetarian'." Published August 5, 2010. Accessed May 15, 2024. https://www.paulmccartney.com/news/if-slaughterhouses-had-glass-walls-everyone-would-be-vegetarian

McGuire, Luke, Sally Palmer, and Nadira S. Faber. "The development of speciesism: Age-related differences in the moral view of animals." *Social Psychological and Personality Science*, 14, no. 2 (2023): 228–237. DOI: https://doi.org/10.1177/194855062210861.

McJetters, Christopher Sebastain. "Slavery. It's still a thing." In *Circles of Compassion: Essays Connecting Issues of Justice*, edited by Will Tuttle, 127–132. Danvers, MA: Vegan Publishers, 2014.

Meijer, Eva, and Bernice Bovenkerk, B. "Taking Animal Perspectives into Account in Animal Ethics." In *The International Library of Environmental, Agricultural and Food Ethics, Volume 33: Animals in Our Midst: The Challenges of Co-existing with Animals in the Anthropocene*, edited by Bernice Bovenkerk and Josef Keulartz, 49–64. Cham, Switzerland: Springer, 2021. DOI: https://doi.org/10.1007/978-3-030-63523-7_3

Meijer, Eva. "Political Communication with Animals." *Humanimalia* 5, no. 1 (2013): 28–52.

Meijer, Eva. *Animal Languages: The Secret Conversations of the Living World*. Translated by L. Watkinson. London: John Murray, 2019a.

Meijer, Eva. *When Animals Speak: Toward An Interspecies Democracy*. New York: New York University Press, 2019b.

Merriam-Webster online dictionary. "anthropocentrism." Published n.d.-a. Accessed May 7, 2024. https://www.merriam-webster.com/dictionary/anthropocentricism.

Merriam-Webster online dictionary. "autonomy." Published n.d.-b. Accessed March 30, 2024. https://www.merriam-webster.com/dictionary/autonomy.

Merriam-Webster online dictionary. "freedom." Published n.d.-c. Accessed March 30, 2024. https://www.merriam-webster.com/dictionary/freedom.

Merriam-Webster online dictionary. "justice." Published n.d.-d. Accessed March 30, 2024. https://www.merriam-webster.com/dictionary/justice.

Milburn, Josh. "Rabbits, Stoats and the Predator Problem: Why a strong animal rights position need not call for human intervention to protect prey from predators." *Res Publica* 1, no. 3 (2015): 273–289. DOI: https://doi.org/10.1007/s11158-015-9281-2.

Milburn, Josh. "Plant sentience and the case for ethical veganism." *Animal Sentience*, 33, no. 5 (2023): 1–2. DOI:https://doi.org/10.51291/2377-7478.1773.

Mill, John Stuart. *Utilitarianism*. London: Parker, Son, and Bourn, 1863.

Mill, John Stuart. (1969) "Three Essays on Religion." In *The Collected Works of John Stuart Mill, Volume X: Essays on Ethics Religion and Society*. Edited by John M. Robson, 184–187. London: Routledge, 1969.

Miller, Suzanne C., Cathy Kennedy, Dale DeVoe, Matthew Hickey, Tracy Nelson, and Lori Kogan. "An examination of changes in oxytocin levels in men and women before and after interaction with a bonded dog." *Anthrozoös* 22 (2009): 31–42. DOI: https://doi.org/10.2752/175303708X390455.

Milligan, Tony. "Putting Pluralism First." In *The Political Turn in Animal Ethics*, edited by Siobhan O'Sullivan and Rober Garner, 85–102. Lanham, MD: Rowman & Littlefield International, 2016.

Monticelli, Lara. "Embodying Alternatives to Capitalism in the 21st Century." *TripleC* 16, no. 2 (2018): 501–17. DOI: https://doi.org/10.31269/triplec.v16i2.1032.

Mora, Camilo, Derek P. Tittensor, Sina Adl, Alastair G.B. Simpson, and Boris Worm. "How Many Species Are There on Earth and in the Ocean?" *PLoS Biology* 9, no. 8 (2011): e1001127. DOI: https://doi.org/10.1371/journal.pbio.1001127.

Morell, Virginia. *Animal Wise: How We Know Animals Think and Feel*. New York: Crown, 2014.

Morris, Christoper. "Justice, Reasons and Moral Standing" In *Rational Commitment and Social Justice: Essays for Gregory Kavka*, edited by J Coleman, and Christopher Morris, 186–207. Cambridge: Cambridge University Press, 1998.

Morton, David B. "Sizeism." In *Encyclopedia of Animal Rights and Animal Welfare*, edited by Marc Bekoff and Carron A. Meaney, 318. Westport, CT: Greenwood Press, 1998.

Mowbray, Alastair. *The Development of Positive Obligations under the European Convention on Human Rights by The European Court of Human Rights*. Oxford: Hart, 2004.

Müller, Nico D. “The 3Rs Alone Will Not Reduce Total Animal Experimentation Numbers: A Fundamental Misunderstanding in Need of Correction.” *Journal of Applied Animal Ethics Research* 5, no. 2 (2023): 269–284. DOI: https://doi.org/10.1163/25889567-bja10042.

Nagasawa, Miho, Takefumi Kikusui, Tatsushi Onaka, and Mitsuaki Ohta. “Dog’s gaze at its owner increases owner’s urinary oxytocin during social interaction.” *Hormones and Behavior* 55 (2009): 434–441. DOI: https://doi.org/10.1016/j.yhbeh.2008.12.002.

National Archives, The. “Farm Animal Welfare Council: The Five Freedoms.” Archived October 10, 2012. https://webarchive.nationalarchives.gov.uk/ukgwa/20121010012427/http://www.fawc.org.uk/freedoms.htm

National Institute of Food and Agriculture (NIFA). “Bioethics.” Published n.d. Accessed February 4, 2015. nifa.usda.gov/nea/animals/pdfs/bioethics_thompson.pdf.

Neuhaus, Winfried, Birgit Reininger-Gutmann, Beate Rinner, Roberto Plasenzotti, Doris Wilflingseder, Joery De Kock, Tamara Vanhaecke, et al. “The Rise of Three Rs Centres and Platforms in Europe.” *Alternatives to Laboratory Animals* 50, no. 2 (2022): 90–120. DOI:https://doi.org/10.1177/02611929221099165.

Ng, Yew-Kwang. “Towards welfare biology: Evolutionary economics of animal consciousness and suffering.” *Biology and Philosophy* 10, no. 4 (1995): 255–285. DOI: https://doi.org/10.1007/BF00852469.

Niesen, Peter. “Landwirtschaftliche Tierhaltung und die Kategorie der Tierarbeit.” *TIERethik* 14, no. 1 (2022): 39–56.

Nonhuman Rights Project (NhRP). “Happy.” Published n.d.-a. Accessed April 19, 2024a. https://www.nonhumanrights.org/client/happy/.

Nonhuman Rights Project (NhRP). “Tommy.” Published n.d.-b. Accessed April 19, 2024b. https://www.nonhumanrights.org/client/tommy/.

Nussbaum, Martha C. *Frontiers of Justice: Disability, Nationality, Species Membership*. Cambridge, MA: Harvard University Press, 2006.

Nussbaum, Martha C. *Frontiers of Justice: Disability, Nationality, Species Membership*. Cambridge, MA: Harvard University Press, 2007.

Nussbaum, Martha C. *Die Grenzen der Gerechtigkeit: Behinderung, Nationalität und Spezieszuhörigkeit*. Translated by Robin Celikates and Eva Engels. Frankfurt am Main: Suhrkamp, 2010.

O’Doherty, Damian P. “Feline politics in organization: The nine lives of Olly the cat.” *Organization*, 23, no. 3 (2016): 407–33. DOI: https://doi.org/10.1177/1350508416629450.

Odendaal, J.S. “Animal-assisted therapy: magic or medicine?” *Journal of Psychosymatic Research* 49 (2000): 275–280. DOI: https://doi.org/10.1016/s0022-3999(00)00183-5.

Odendaal, J.S., and R.A. Meintjes. "Neurophysiological correlates of affiliative behaviour between humans and dogs." *Veterinary Journal* 165 (2003) 296–301. DOI: https://doi.org/10.1016/s1090-0233(02)00237-x

Our Planet. Theirs Too. "The Declaration of Animal Rights." Last modified 2011. Accessed April 10, 2024. https://declarationofar.org/.

Oxford Cenre for Animal Ethics. "Ethics of Animal Experimentation." Published n.d. Accessed February 2, 2023. https://www.oxfordanimalethics.com/what-we-do/ethics-of-animal-experimentation/.

Pachirat, Timothy. *Every Twelve Seconds: Industrialized Slaughter and the Politics of Sight*. New Haven, CT: Yale University Press, 2011.

Pacini-Ketchabaw, Veronica, Mindy Blaise, and Affrica Taylor. "De-centring the human in multispecies ethnographies." In *Posthuman Research Practices*, edited by Carol Taylor and Christina Hughes, 149–167. Basingstoke: Palgrave Macmillan, 2016.

Palmer, Clare. "Animal Disenhancement and the Non-Identity Problem: A Response to Thompson." *Nanoethics* 5 (2011): 43–48. DOI: https://doi.org/10.1007/s11569-011-0115-1.

Pasquet, Vivian. "Wenn jede Minute zur Qual wird." *Zeit Online*. Published January 15, 2015. Accessed November 27, 2015. zeit.de/2015/03/contergan-medizin-arzt-behinderung.

Pauen, Michael. *Was ist der Mensch? Die Entdeckung der Natur des Geistes*. Munich: DVA, 2007.

Pauketat, Janet and Jacy Anthis. "Predicting the moral consideration of artificial intelligences." *Computers in Human Behavior*. 136 (2022): 107372. DOI: https://doi.org/10.1016/j.chb.2022.107372.

Pearce, David. *The Hedonistic Imperative (1995)*. Accessed March 23, 2020. http://www.hedweb.com/hedethic/tabconhi.htm.

Pearce, David. "Reprogramming predators." Published 2009. Last modified 2015. Accessed March 23, 2020. http://www.hedweb.com/abolitionist-project/reprogramming-predators.html.

People for the Ethical Treatement of Animals (PETA). "In-vitro-Fleisch und Clean Meat: Gibt es Fleisch aus dem Labor?" Last modified January 13, 2019. Accessed February 16, 2015. peta.de/laborfleisch#.VOI4wy6umqE.

People for the Ethical Treatment of Animals (PETA). "Bigger Fish to Free: PETA's Complete List of Animal-Friendly Idioms." Last modified n.d.-a. Accessed February 17, 2024. https://www.peta.org/features/animal-friendly-idioms/.

People for the Ethical Treatment of Animals (PETA). "Entenmast." Last modified n.d.-b. Accessed November 7, 2015. peta.de/entenmast#.Vj5blpUrmUk.

Pepper, Angie. "Is Animal Labour a Viable Route to Interspecies Justice?" *Politics and Animals* 8 (2022): 12–17. DOI: https://doi.org/10.1111/japp.12436.

Perel, Pablo, Ian Roberts, Emily Sena, Philipa Wheble, Catherine Briscoe, Peter Sandercock, Malcolm Macleod, et al. "Comparison of treatment effects between animal experiments and clinical trials: systematic review." *British Medical Journal* 334 (2007): 197–203. DOI: https://doi.org/10.1136/bmj.39048.407928.BE.

Pianka, Eric R. "On r- and K-Selection." *American Naturalist* 104 (1970): 592–597 Pleasance, Chris. "Police Dogs to get a Pension Plan." *The Daily Mail.* Last modified November 4, 2013. https://www.dailymail.co.uk/news/article-2487540/Police-dogs-pension-plan-Animals-given-1-500-help-pay-medical-bills-retire-service.html

Plumwood, Val. *Feminism and the Mastery of Nature.* London: Routledge, 1993. Plumwood, Val. "Human Vulnerability and the Experience of Being Prey." *Quadrant* 39, no. 3 (March 1995): 29–34.

Plumwood, Val. *Environmental Culture: The Ecological Crisis of Reason.* London: Routledge, 2002.

Plutarch. *Fünf Doppelbiographien: Teil 1: Alexandros und Caesar. Aristeides und Marcus Cato. Perikles und Fabius Maximus. Teil 2: Gaius Marius und Alkibiades. Demosthenes und Cicero.* Translated by Konrat Ziegler and Walter Wuhrmann. Berlin: De Gruyter, 2001.

Porcher, Jocelyne. *The Ethics of Animal Labor: A Collaborative Utopia.* Cham, Switzerland: Palgrave Macmillan, 2017

Porcher, Jocelyne, and Jean Estebanez. "Animal Labor. At the Forefront of Innovative Research." In *Animal Labor: A New Perspective on Human-Animal Relations*, edited by Jocelyne Porcher, and Jean Estebanez, 11–33. Bielefeld: Transcript, 2019.

Post, Mark J. "Interview with Mark Post." Accessed October 7, 2015. static.orf.at/podcast/fm4/fm4_reality_check_podcast.xml.

Post, Mark J., Shulamit Levenberg, David L. Kaplan, Nicholas Genovese, Jianan Fu, Chris Bryant, Nicole Negowetti, et al. "Scientific, sustainability and regulatory challenges of cultured meat." *Nature Food* 1, no. 7 (2020): 403–415. DOI: https://doi.org/10.1038/s43016-020-0112-z.

Precht, Richard David. *Noahs Erbe – Vom Recht der Tiere und den Grenzen des Menschen.* Reinbek bei Hamburg: Rowohlt Taschenbuch Verlag, 2000.

Precht, Richard David. *Liebe – ein unordentliches Gefühl.* Munich: Goldmann, 2009.

Precht, Richard David. *Warum gibt es alles und nicht nichts?* Munich: Goldmann, 2011.

Pro Familia. *Sexualität und geistige Behinderung.* Frankfurt am Main: Pro Familia, 2006.

Rachels, James. "Do Animals Have a Right to Liberty?" In *Animal Rights and Human Obligations*, edited by Tom Regan, and Peter Singer, 205–223. Englewood Cliffs, NJ: Prentice Hall, 1976.

Rechtsinformationssystem des Bundes (RIS). "Bundesrecht konsolidiert: Gesamte Rechtsvorschrift für Verbotsgesetz 1947." Published n.d.-a. Accessed May 16, 2024. https://www.ris.bka.gv.at/GeltendeFassung.wxe?Abfrage=Bundesnormen&Gesetzesnummer=10000207.

Rechtsinformationssystem des Bundes (RIS). "Bundesrecht konsolidiert: Gesamte Rechtsvorschrift für 2. Tierhaltungsverordnung." Published n.d.-b. Accessed May 20, 2024. https://ris.bka.gv.at/GeltendeFassung.wxe?Abfrage=Bundesnormen&Gesetzesnummer=20003860.

Rechtsinformationssystem des Bundes (RIS). "Bundesrecht konsolidiert: Gesamte Rechtsvorschrift für Tierschutzgesetz." Published n.d.-c. Accessed May 20, 2024. https://ris.bka.gv.at/GeltendeFassung.wxe?Abfrage=Bundesnormen&Gesetzesnummer=20003541.

Rechtsinformationssystem des Bundes (RIS). "Bundesrecht konsolidiert: Gesamte Rechtsvorschrift für Tierversuchsgesetz 2012." Published n.d.-d. Accessed May 21, 2024. https://ris.bka.gv.at/GeltendeFassung.wxe?Abfrage=Bundesnormen&Gesetzesnummer=20008142.

Rechtsinformationssystem des Bundes (RIS). "Bundesrecht kondolidiert: Artikel § 285a." Published n.d.-e. Accessed April 22, 2016. https://www.ris.bka.gv.at/Dokument.wxe?Abfrage=Bundesnormen&Dokumentnummer=NOR12018011.

Rechtsinformationssystem des Bundes (RIS). "Bundesrecht konsolidiert: Artikel §§ 36 & 37. Published n.d.-f. Accessed November 26, 2015. https://ris.bka.gv.at/GeltendeFassung.wxe?Abfrage=Bundesnormen&GesGesetzesnum=20003541.

Regan, Tom. The Case for Animal Rights. Berkeley, CA: University of California Press, 1983.

Regan, Tom. *The Case for Animal Rights.* London: Routledge, 1984.

Regan, Tom. *The Case for Animal Rights.* Berkeley, CA: University of California Press: 2004.

Regan, Tom. *Defending Animal Rights.* Chicago: University of Illinois Press, 2006.

Regan, Tom. Interview on DVD *Live and Let Live.* Film directed by Marc Pierschel. Mindjazz Pictures, 2014.

Reinecke, Juliane. "Social movements and prefigurative organizing: Confronting entrenched inequalities in Occupy London." *Organization Studies* 39, no. 9 (2018): 1299–1321. DOI: https://doi.org/10.1177/0170840618759815.

Rigato, Emanuele, and Alessandro Minelli. "The great chain of being is still here." *Evolution: Education and Outreach* 6, no. 18 (June 2013): 1–6. DOI:https://doi.org/10.1186/1936-6434-6-18.

Rößinger, Monika. „Bedrohte Seepferdchen." Last modified January 1, 2013. Accessed February 2, 2015. spektrum.de/alias/zoologie/bedrohteseepferdchen/830394?_ga=1.2049718.602498648.422914275.

Robinson, M. "Veganism and Mi'kmaq Legends." Journal of Native Studies 33, no. 1 (2013): 189–196.

Rodogno, Raffaele. "Sentientism, wellbeing, and environmentalism." *Journal of Applied Philosophy* 27, no. 1 (2010): 84–99. DOI: https://doi.org/10.1111/j.1468-5930.2009.00475.x.

Rollin, Bernard E. "The Ascent of Apes – Broadening the Moral Community." In *The Great Ape Project – Equality Beyond Humanity*, edited by Paola Cavalieri and Peter Singer, 206–219. New York: St. Martin's Press, 1993.

Rollin, Bernard E. *Animal Rights & Human Morality*. Amherst, NY: Prometheus Books, 2006.

Rollin, Bernard E. *Putting the Horse before Descartes: My Life's Work on Behalf of Animals*. Philadelphia: Temple University Press, 2011.

Román-Palacios, Cristian, and Wiens, John J. "Recent responses to climate change reveal the drivers of species extinction and survival." *Proceedings of the National Academy of Sciences of the United States of America*. 117, no. 8 (2020): 4211–4217. DOI: https://doi.org/10.1073/pnas.1913007117.

Rorheim, Adrian, Adriano Mannino, Tobias Baumann, and Lucius Caviola. "Cultured Meat: A pragmatic solution to the problems posed by industrial farming: Policy paper." *Sentience Politics* 1 (2016): 1–14. https://ea-foundation.org/files/cultured-meat.pdf.

Rothaug, Martin. "Sexuelle Selbstbestimmung bei schwerer Behinderung." In *Sexualität leben ohne Behinderung: Das Menschenrecht auf sexuelle Selbstbestimmung*, edited by Jens Clausen, and Frank Herrath, 47–71. Stuttgart: Kohlhammer, 2013.

Rowlands, Mark. "Animal Rights." In *The International Encyclopedia of Ethics*, edited by Hugh LaFollette, Hoboken NJ: Wiley-Blackwell, 2013. (Penultimate version. Accessed October 28, 2015. academia.edu/17280516/Animal_Rights.

Roy, Arundhati. "The 2004 Sydney Peace Prize Lecture: Peace and the New Corporate Liberation Theology." Seymour Theatre Centre, University of Sydney. November 4, 2004. https://sydneypeacefoundation.org.au/peace-prize-recipients/2004-arundhati-roy/.

Ryder, Richard D. *Animal Revolution: Changing Attitudes Towards Speciesism*. Oxford: Berg, 2000.

Salomon, Daniel. "Animal Rights and Autism Pride: Let's Heal the Rift." *The Scavenger*, October 2010.

Salt, Henry S. *Animals' Rights: Considered in Relation to Social Progress*. New York: Macmillan & Co., 1894.

Sapontzis, Steve F. "Predation." *Ethics and Animals* 5, no. 2 (1984): 27–38. DOI: https://doi.org/10.15368/ea.1984v5n2.1.

Sapontzis, Steve F. *Morals, Reason, and Animals*. Philadelphia: Temple University Press, 1987.

Sayers, J. G. "A report to an academy: On carnophallogocentrism, pigs and meat-writing." *Organization* 23, no. 3 (2016): 370–386. DOI: https://doi.org/10.1177/1350508416629454.

schaf-foren.org: Fachforum für die Haltung und Zucht von Schafen und die Verarbeitung von Produkten der Schafhaltung. Published n.d. Accessed May 2, 2016. schaf-foren.de/forum/index.php?showtopic=17127&st=25.

Schirnhofer, Heinz. “Fisch + Fleisch: Schweineskandal – Stecken wir nicht alle im Boot?” Published November 21, 2015. Accessed November 25, 2015. https://www.fischundfleisch.com/styrias-best/schweineskandal-stecken-wir-nicht-alle-im-boot-12053.

Scheele, Dirk, Andrea Wille, Keith M. Kendrick, Birgit Stoffel-Wagner, Benjamin Becker, Onur Güntürkün, Wolfgang Maier, and Rene Hurleman. “Oxytocin enhances brain reward system responses in men viewing the face of their female partner.” *Proceedings of the National Academy of Sciences of the United States of America* 110.50 (2013): 20308–20313. DOI: https://doi.org/10.1073/pnas.1314190110.

Schlosberg, David. *Defining Environmental Justice: Theories, Movements and Nature*. Oxford: Oxford University Press, 2007.

Schlosberg, David. “Ecological Justice for the Anthropocene.” In *Political Animals and Animal Politics*, edited by Marcel Wissenburg, and David Schlosberg, 75–89. London: Palgrave Macmillan, 2014.

Schmidt, Veronika. “Ganz normal: ‘Homo-Ehe’ im Tierreich.” Published April 15, 2008. Accessed December 27, 2014. diepresse.com/home/science/376626/Ganz-normal_HomoEhe-im-Tierreich.

Schneeberger, Doris. “Nichtmenschliche Tiere in Organisationen – Produkte, Sklav*innen, oder Arbeiter*innen?” *TIERethik* 14, no. 1 (2022): 57–83.

Schopenhauer, Arthur. “Preisschrift über die Grundlage der Moral” in *Werke in zehn Bänden*: *Vol. VI*, 143–317. Edited by Arthur Hübscher Zurich: Zurich Edition, 1977.

Schopenhauer, Arthur. *Die Welt als Wille und Vorstellung*. Paderborn: Voltmedia, 2005.

Schopenhauer, Arthur. *Senilia – Gedanken im Alter*. Edited by Franco Volpi and Ernst Ziegler. Munich: Beck, 2010.

Schopenhauer, Arthur. *Aphorismen zur Lebensweisheit*. Published n.d. Accessed August 16, 2015. dnb.info/1041219032/34.

Schulze, Wilhelm, H.S. Schultze-Petzold, A.S. Hazem, and R. Gross. “Versuche zur Objektivierung von Schmerz und Bewußtsein bei der konventionellen (Bolzenschußbetäubung) sowie religionsgesetzlichen (‘Schächtschnitt’) Schlachtung von Schaf und Kalb.” *Deutsche Tierärztliche Wochenschrift* 85, no. 2 (1978): 62–66.

Schwarz, Peter, and Monika Berger-Lenz. *All About Mini Pigs*. Norderstedt: Books on Demand, 2002.

Schweitzer, Albert. *Die Ehrfurcht vor dem Leben*. Munich: C.H. Beck, 2008.

Schweppenhäuser, Gerd. *Die Antinomie des Universalismus: Zum moralphilosophischen Diskurs der Moderne*. Würzburg: Königshausen und Neumann, 2005.

Sedmak, Clemens. *Geglücktes Leben*. Freiburg: Herder, 2012.

Seeley, Tom D. *Honeybee Democracy*. Princeton NJ: Princeton University Press, 2010.

Sen, Amartya. "Development as Capability Expansion." *Journal of Development Planning* 19, no. 1 (1989): 41–58.

Sen, Amartya. "Capability and Well-Being." In *The Quality of Life*, edited by Martha Nussbaum, and Amartya Sen, 30–53. New York: Routledge, 2004.

Sezgin, Hilal. *Artgerecht ist nur die Freiheit.* Munich: C.H. Beck, 2014a.

Sezgin, Hilal. *Hilal Sezgins Tierleben: Von Schweinen und anderen Zeitgenossen.* Munich: C.H. Beck, 2014b.

Shafran, Avi. "This Time PETA's Guilty of Missing the Point." *Jewish News Weekly of Northern California*. Published May 20, 2005. http://www.jewishsf.com/content/20/module/displaystory/story_id/25916/format/html/displaystory.html.

Shaw, Rosemary. "A Case for Recognizing the Rights of Animals as Workers." *Journal of Animal Ethics* 8, no. 2 (2018): 182–198. DOI: https://doi.org/10.5406/janimalethics.8.2.0182.

Shepherdson, David J. "Tracing the Path of Environmental Enrichment in Zoos." In *Second Nature: Environmental Enrichment for Captive Animals*, edited by David J. Shepherdson, Jill D. Mellen, and Michael Hutchins, 1–14. Washington: Smithsonian Institution Press, 1998.

Shoemaker, Sydney. "Persons, Animals, and Identity." *Synthesis* 162, no. 3 (2008): 313–324. DOI: https://doi.org/10.1007/s11229-007-9253-y.

Shue, Henry. *Basic Rights: Subsistence, Affluence, and US Foreign Policy*. Princeton, NJ: Princeton University Press, 1996.

Sielmann Stiftung. "Wiederansiedlung von Przewalski-Pferden." Published n.d. Accessed October 11, 2015. sielmann-stiftung.de/ausdruck/projekte/projekte-international/wiederansiedlung-von-przewalski-pferden/.

Sikimic, Vlasta, and Andrea Berber. "On the Moral and Legal Status of Animals." Accessed September 9, 2015. academia.edu/13891072/Andrea_Berber_and_Vlasta_Sikimic_On_the_Moral_and_Legal_Status_of_Animals.

Simmons, Aaron. "Animals, predators, the right to life and the duty to save lives." *Ethics and the Environment* 14 (2009): 15–27. DOI: https://doi.org/10.2979/ete.2009.14.1.15.

Singer, Peter. *Animal Liberation*. New York: Harper Collins, 1975.

Singer, Peter. *Practical Ethics*. Cambridge: Cambridge University Press, 1979.

Singer, Peter. *Praktische Ethik*. Translated by Oscar Bischoff, Jean-Claude Wolf, Dietrich Klose, and Susanne Lenz. Stuttgart: Reclam, 1994.

Singer, Peter. "Reply to Martha Nussbaum, Justice for Nonhuman Animals, The Tanner Lectures on Human Values." Published November 13, 2002. Accessed June 15, 2015. utilitarian.net/singer/by/20021113.htm.

Singer, Peter. *Animal Liberation*. New York: Harper Collins, 2009.

Singer, Peter. *The Expanding Circle*. Princeton, NJ: Princeton University Press, 2011.

Singer, Peter. *The Most Good You Can Do: How effective altruism is changing ideas about living ethically*. New Haven, CT: Yale University Press, 2016

Smart, Alan. "Critical perspectives on multispecies ethnography." *Critique of Anthropology* 34, no. 1 (2014): 3–7. DOI: https://doi.org/10.1177/0308275X13510749.

Smith, Christie. "Articulating Ecological Injustices of Recognition." In *Political Animals and Animal Politics*, edited by Marcel Wissenburg and David Schlosberg, 61–74. London: Palgrave Macmillan, 2014.

Solar Foods. "Solein: Protein out of thin air." Published 2023. Accessed May 7, 2024. https://www.solein.com.

Spannring, Reingard. "Mutual Becomings? In Search of an Ethical Pedagogic Space in Human-Horse Relationships: Interdisciplinary Approaches to Curriculum and Pedagogy." In *Animals in Environmental Education*, edited by Teresa Lloro-Bidart, and Valerie S. Banschbach, 79–94. Cham, Switzerland: Palgrave Macmillan. DOI: https://doi.org/10.1007/978-3-319-98479-7_5.

Spiegel, Der. "Regierung rügt Tierquälerei in Schlachthöfen." Published June 21, 2012. Accessed November 10, 2015. spiegel.de/wissenschaft/natur/schlachthoefe-arbeiten-mit-hoher-fehlerquote-tiere-leiden-unnoetig-a-840156.html.

Stegmüller-Lang, Sabine. *Tierschutz macht Sinn*. Neckenmarkt: novum pro, 2014.

Stephen, Leslie. *Social Rights and Duties: Addresses to Ethical Societies. Volume I*. London: Swan Sonnenschein & Company, 1896.

Stone, Christopher D. *Should Trees Have Standing? Law, Morality, and the Environment*. Oxford: Oxford University Press, 2010.

Strasser, Hiltrud. Lecture entitled "Can Animal Experiments in the Laboratory be Scientific?" University of Salzburg. November 26, 2015.

Stucki, Saskia. *One Rights: Human and Animal Rights in the Anthropocene*. Cham, Switzerland: Springer, 2023.

Szabo, Birgit, and Eva Ringler. "Geckos differentiate self from other using both skin and faecal chemicals: evidence towards self-recognition?" *Animal Cognition* 26, no. 3 (2023): 1011–1019. DOI: https://doi.org/10.1007/s10071-023-01751-8.

Tanasescu, Mihnea. (2014) "The Rights of Nature: Theory and Practice." In *Political Animals and Animal Politics*, edited by Marcel Wissenburg and David Schlosberg, 150–163. London: Palgrave Macmillan, 2014.

Taylor, Angus. "An Interview with Sue Donaldson and Will Kymlicka." *Between the Species* 17, no. 1 (2014): 140–165. DOI: https://doi.org/10.15368/bts.2014v17n1.7.

Taylor, Nik, and Heather Fraser. "The Work of Significant Other/s: Companion Animal Relationships in the Future." In *Companion Animals and Domestic Violence*, edited by Nik Taylor, and Heather Fraser, 185–217. Cham, Switzerland: Palgrave Macmillan, 2019.

Taylor, Sunaura. *Beasts of Burden: Animal and Disability Liberation*. New York: The New Press, 2017.

Tageszeitung, Die (*taz*). "Jagd auf Delfine in Japann – Das Land des Gemetzels." Published September 16, 2014. Accessed February 15, 2015. taz.de/Jagd-auf-Delfine-in-Japan/!146071.

Tageszeitung, Die (*taz.*). "Japanische Fangschiffe vor der Antarktis – Auf Waljagd ohne Waffen." Published January 9, 2015. Accessed February 15, 2015. https://taz.de/Japanische-Fangschiffe-vor-der-Antarktis/!5024301/.

Terrill, Ethan C. "Plants, partial Moral Status, and practical Ethics." *Journal of Consciousness Studies* 28. No. 1–2 (2021): 184–209.

Terrill, Ethan C., and Walter Veit. "Better to be a Pig Dissatisfied than a Plant Satisfied." *Journal of Agricultural and Environmental Ethics* 37, no. 4 (2024):1–17. DOI: https://doi.org/10.1007/s10806-024-09922-4.

Thompson, Paul B. "The Opposite of Human Enhancement: Nanotechnology and the Blind Chicken Problem." *Nanoethics* 2 (2008): 305–316. DOI: https://doi.org/10.1007/s11569-008-0052-9

Tierschutz Ombudsstelle. "Der vernünftige Grund für die Tötung von Tieren." Published n.d. Accessed September 24, 2015. https://www.tierschutzombudsstelle.steiermark.at/cms/dokumente/11642258_50074656/2f4242b6/Tierärztekammer_Beitrag_Der%20vernünftige%20Grund%20für%20die%20Tötung%20von%20Tieren.pdf

Tomasik, Brian. "The Importance of Wild-Animal Suffering." *Relations Beyond Anthropocentrism* 3, no. 2 (2015): 133–152. DOI: https://doi.org/10.7358/rela-2015-002-toma.

Toni's Freilandeier. "Henne und Hahn." Published n.d. Accessed September 24, 2015. www.tonishennehahn.at.

Towne Hirtenfelder, Claudia, and Carolyn Prouse. "Milking economies: Multispecies entanglements in the infant formula industry." *Environment and Planning E: Nature and Space*, 5, no. 3 (2022): 1296–1318. DOI: https://doi.org/10.1177/25148486211036.

Tracey, Daniel W., Rachel Wilson, Gilles Laurent, and Seymour Benzer. "painless, a Drosophila Gene Essential for Nociception." *Cell* 113, no. 2 (2003): 261–273. DOI: https://doi.org/10.1016/s0092-8674(03)00272-1.

Tuttle, Will. "Introduction: The Circles of Compassion Vision." *Circles of Compassion: Essays Connecting Issues of Justice*, edited by Will Tuttle, 19–21. Danvers, MA: Vegan Publishers, 2014.

UN-Behindertenrechtskonvention. "Meinungsfreiheit und Informationszugang." Published n.d. Accessed November 4, 2015. behindertenrechtskonvention.info/meinungsfreiheit-und-informationszugang-3879.

United Nations. "Nothing about Us, Without Us." Published 2004. Accessed March 26, 2024. https://www.un.org/esa/socdev/enable/iddp2004.htm#:~:text=The%20motto%20%E2%80%9CNothing%20About%20Us,and%20with%20persons%20with%20disabilities

United Nations. "General Assembly Adopts Groundbreaking Convention, Optional Protocol on Rights of Persons with Disabilities." Press release. New York: United Nations, December 13, 2006.

United Nations Foundation. "The Universal Declaration of Human Rights turning 75: What you need to know." Published December 6, 2023. Accessed June 05, 2024. https://unfoundation.org/blog/post/the-universal-declaration-of-human-rights-is-turning-75-heres-what-you-need-to-know/.

United Nations Children's Fund (UNICEF). "Convention on the Rights of the Child." Published n.d. Accessed March 26, 2024. https://www.unicef.org/child-rights-convention/convention-text.

United Nations Treaty Collection. "Chapter IV: Human Rights. 4. International Covenant on Civil and Political Rights." Published n.d.-a. Accessed March 17, 2024. https://treaties.un.org/Pages/ViewDetails.aspx?src=TREATY&mtdsg_no=IV-4&chapter=4&clang=_en.

United Nations Treaty Collection. "Chapter IV: Human Rights. 3. International Covenant on Economic, Social and Cultural Rights." Published n.d.-b. Accessed March 17, 2024. https://treaties.un.org/Pages/ViewDetails.aspx?src=IND&mtdsg_no=IV-3&chapter=4.

United Poultry Concerns. "The Experimental Use of Chickens and Other Birds in Biomedical and Agricultural Research: Conclusion – Night of the Living Dead for Birds." Published n.d. Accessed February 4, 2015. upc-online.org/experimentation/experimentalConclusion.htm.

United States National Archives and Records Administration. "America's Founding Documents." Last modified June 5, 2023. Accessed August 12, 2015. https://www.archives.gov/founding-docs.

Universal Declaration of Human Rights. Approved and proclaimed by the General Assembly of the United Nations in 1948. Accessed November 1, 2015. un.org/depts/german/menschenrechte/aemr.pdf.

Vaillant, George E. "Happiness is Love: Full Stop". Unpublished manuscript, 2012. Accessed November 20, 2024. https://www.duodecim.fi/xmedia/duo/pilli/duo99210x.pdf.

van Elderen, Heidi. "Bei manchen Tieren hält die Liebe ein Leben lang." Last modified February 14, 2013. Accessed December 27, 2014. tierwelt.ch/?rub=4503&id=34463.

van Roojen, Mark. "Moral Cognitivism vs. Non-Cognitivism." In *The Stanford Encyclopedia of Philosophy Archive: Winter 2023 Edition*, edited by Edward N. Zalta, and Uri Nodelman. Last modified 2023. https://plato.stanford.edu/archives/win2023/entries/moral-cognitivism/.

Varkey, Basil. "Principles of Clinical Ethics and Their Application to Practice." *Medical Principles and Practice* 30, no. 1 (2021): 17–28. https://doi.org/10.1159/000509119.

Vine. "About us." Last modified 2018. http://vine.bravebirds.org/about-us/.

Višak, Tatjana. "Do Utilitarians Need to Accept the Replaceability Argument?" In *The Ethics of* Killing Animals, edited by Tatjana Višak and Robert Garner, 117–135. Oxford: Oxford University Press, 2015.

Volkelt, Johannes. *Arthur Schopenhauer*. Paderborn: Salzwasser Verlag, 2014.

Wachowski, Lana, Lilly Wachowski, and Tom Tykwer, directors. *Cloud Atlas*. Cloud Atlas Productions, X Filme Creative Pool, and Anarchos Productions. 2012.

Wadiwel, Dinesh J. *The War Against Animals*. Leyden, The Netherlands: Brill, 2015.

Wadiwel, Dinesh J. "The Working Day: Animals, Capitalism, and Surplus Time." In *Animal Labour: A New Frontier of Interspecies Justice?* edited by Charlotte E. Blattner, Kendra Coulter, and Will Kymlicka, 181–206. Oxford: Oxford University Press, 2020.

Wallace, R. Jay. "Practical Reason." In *The Stanford Encyclopedia of Philosophy Archive: Summer 2009 Edition*, edited by Edward N. Zalta and Uri Nodelman. Last modified 2009. http://plato.stanford.edu/archives/sum2009/entries/practical-reason/.

walrecht de. "Karsten Brensing gemeinsam mit Veronica Ferres umd Alicia von Ritterberg bei Markus Lanz." Uploaded to YouTube April 23, 2013a. Accessed November 19, 2015a. youtube.com/watch?v=caiMyMZBoj0.

walrecht de. "Karsten Brensing gemeinsam mit Veronica Ferres umd Alicia von Ritterberg bei Markus Lanz 2/2." Uploaded to YouTube April 23, 2013b. Accessed November 19, 2015b. https://www.youtube.com/watch?v=_xogvPhDWYs.

walrecht de. "Bundestagsanhörung Delfine Mai 2013." Uploaded to YouTube September 24, 2013. Accessed October 2, 2015. youtube.com/watch?v=JxkjjNB1od0>

Weeks, Kathi. *The Problem with Work: Feminism, Marxism, Antiwork Politics, and Postwork Imaginaries*. Durham NC: Duke University Press, 2011.

Weil, Kari. *Thinking Animals: Why Animal Studies Now?* New York: Columbia University Press, 2012.

Welt. "Cecils Jäger glaubt, alles richtig gemacht zu haben." Published July 29, 2015. Accessed October 6, 2015. welt.de/vermischtes/weltgeschehen/article144565352/Cecils-Jaeger-glaubt-alles-richtig-gemacht-zu-haben.html.

Wibye, Johan V. "Reviving the Distinction between Positive and Negative Human Rights." *Ratio Juris* 35 (2022): 363–382. DOI: https://doi.org/10.1111/raju.12363.

Wilford, John Noble. "Almost Human, and Sometimes Smarter." Published April 17, 2007. Accessed January 6, 2015. nytimes.com/2007/04/17/science/17chimp.html?pagewanted=print&_r=0.

Wilks, Matti., Lucius Caviola, Guy Kahane, and Paul Bloom. "Children prioritize humans over animals less than adults do." *Psychological Science* 32, no. 1 (2021): 27–38. DOI: https://doi.org/10.1177/0146167223121939l.

Willett, Walter, Johan Rockström, Brent Loken, Marco Springmann, Tim Lang, Sonja Vermeulen, Tara Garnett, et al. "Food in the Anthropocene: the EAT-Lancet Commission on healthy diets from sustainable food systems." *The Lancet Comissions* 393 (10170) (2019): 447–492. DOI: https://doi.org/10.1016/S0140-6736(18)31788-4.

Wise, Steven M. *Though the Heavens May Fall: The Landmark Trial That Led to the End of Human Slavery*. Cambridge, MA: Perseus Books, 2005.

Wissenburg, Marcel. "An Agenda for Animal Political Theory." In *Political Animals and Animal Politics*, edited by Marcel Wissenburg, and David Schlosberg, 30-43. London: Palgrave Macmillan, 2014.

Wissenburg, Marcel, and David Schlosburg. *Political Animals and Animal Politics*. London: Palgrave Macmillan, 2014.

Wittgenstein, Ludwig. *Tractatus logico-philosophicus*. Translated by C. K. Ogden. Introduced by Bertrand Russell. London: Routledge & Kegan Paul, 1922.

Wohlleben, Peter. *Die Gefühle der Tiere*. Darmstadt: pala-verlag, 2014.

Wolfe, Cary. Animal Rites: *American Culture, the Discourse of Species, and Posthumanist Theory*. Chicago: The University of Chicago Press, 2003.

Wolf, Jean-Claude. "Singer über Rechte, Recht auf Leben und Euthanasie." *Analyse & Kritik* 12 (1990): 219–225. DOI: https://doi.org/10.1515/auk-1990-0208.

Wolf, Jean-Claude. "Kommunitarismus versus Liberalismus." In *Kommunitarismus versus Liberalismus*, edited by Kurt Seelmann, 37–60. Stuttgart: Franz Steiner Verlag, 2000.

Wolfe, Cary. "Turning Toward and Away." In *Messy Eating: Conversations on Animals as Food*, edited by Samantha King, R. Scott Carey, Isabel Macquarrie, Victoria Niva Millious, and Elaine M. Power, 19–35. London: Fordham University Press, 2019.

Woodard, Vincent. *The Delectable Negro: Human Consumption and Homoeroticisim within US Slave Culture*. New York: New York University Press, 2014.

World Society for the Protection of Animals (WSPA; now World Animal Protection). "Universal Declaration of Animal Welfare." Published 2007. Accessed February 14, 2023. worldanimalprotection.ca/sites/default/files/media/ca__en_files/case_for_a_udaw_tcm22-8305.pdf.

World Wide Fund for Nature (WWF). "Species: Lion – Facts." Published n.d. Accessed May 7, 2024. https://www.worldwildlife.org/species/lion%2D%2D19

Wuketits, Franz M. "Hilft es den Tieren, wenn wir Menschen uns selbst als Tiere wahrnehmen?" *TIERethik* 1 (2015) 11–24.

Young, Larry J., Miranda M. Lim., Brenden Gingrich, and Thomas R. Insel. "Cellular mechanisms of social attachment." *Hormones and Behavior* 40, no. 2 (2011): 133–138. DOI: https://doi.org/10.1006/hbeh.2001.1691.

Young, Larry J., Zuoxin Wang, and Thomas R. Insel. "Neuroendocrine bases of monogamy." *Trends in Neuroscience* 21, no. 2 (1998): 71–75. DOI: https://doi.org/10.1016/s0166-2236(97)01167-3.

Zeit Online. Accessed October 13, 2015. zeit.de/politik/deutschland/2015-09/day-old-chicks-bundesrat-toetung-verbot?print=true#comments.

Zeit Online. "Agrarminister will Kükenschreddern beenden." Published July 9, 2015. Accessed October 13, 2015. zeit.de/wirtschaft/2015-07/kueken-schreddern-landwirtschaftsminister-schmidt-in-ovo

Zeit Online. "Kükentöten wird ab 2022 verboten." Last modified January 20, 2021. Accessed November 7, 2022. https://www.zeit.de/politik/deutschland/2021-01/tierschutz-kueken-toeten-verbot-gesetzentwurf-bundeskabinett-agrarpolitik-landwirtschaft.

Zinsmeister, Julia. "Rechtsfragen der Sexualität, Partnerschaft und Familienplanung." In *Sexualität leben ohne Behinderung: Das Menschenrecht auf sexuelle Selbstbestimmung*, edited by Jens Clause, and Frank Herrath, 147–159. Stuttgart: Kohlhammer, 2013.

Index[1]

[1] Note: Page numbers followed by 'n' refer to notes.

D. Schneeberger, *Envisioning a Better Future for Nonhuman Animals*, The Palgrave Macmillan Animal Ethics Series,
https://doi.org/10.1007/978-3-031-75332-9

GPSR Compliance
The European Union's (EU) General Product Safety Regulation (GPSR) is a set of rules that requires consumer products to be safe and our obligations to ensure this.

If you have any concerns about our products, you can contact us on

ProductSafety@springernature.com

In case Publisher is established outside the EU, the EU authorized representative is:

Springer Nature Customer Service Center GmbH
Europaplatz 3
69115 Heidelberg, Germany

www.ingramcontent.com/pod-product-compliance
Lightning Source LLC
Chambersburg PA
CBHW070010020626
46050CB00004B/66
* 9 7 8 3 0 3 1 7 5 3 3 1 2 *